SECOND EDITION

FOOD
IRRADIATION
A Guidebook

SECOND EDITION

FOOD IRRADIATION

A Guidebook

Morton Satin

CRC Press
Taylor & Francis Group
Boca Raton London New York

CRC Press is an imprint of the
Taylor & Francis Group, an **informa** business

CRC Press
Taylor & Francis Group
6000 Broken Sound Parkway NW, Suite 300
Boca Raton, FL 33487-2742

First issued in hardback 2017

CRC Press is an imprint of Taylor & Francis Group, an Informa business

Library of Congress Cataloging-in-Publication Data

Main entry under title:
 Food Irradiation: a Guidebook, Second Edition
                    Full Catalog record is available from the Library of Congress

© 1996 by CRC Press LLC
Originally published by Technomic Publishing

ISBN 13: 978-1-138-42659-7 (hbk)
ISBN 13: 978-1-56676-344-8 (pbk)

No claim to original U.S. Government works
Library of Congress Card Number 95-61619

In loving memory of Tilly, Arthur, Percy and Mamie

Young men, have confidence in those powerful and safe methods, of which we do not yet know all the secrets. And, whatever your career may be, do not let yourselves become tainted by a deprecating and barren skepticism, do not let yourselves be discouraged by the sadness of certain hours which pass over nations. Live in the serene peace of laboratories and libraries. Say to yourselves first: "What have I done for my instruction?" and, as you gradually advance, "What have I done for my country?" until the time comes when you may have the immense happiness of thinking that you have contributed in some way to the progress and to the good of humanity. But, whether our efforts are or not favored by life, let us be able to say, when we come near the great goal, "I have done what I could."

Louis Pasteur
December 27, 1892

CONTENTS

Although the interlude has been rather brief, a great deal of relevant activity has taken place since the first edition of *Food Irradiation: A Guidebook* was published. Two incidents that were highlighted in the final chapter of the original book have blossomed into landmark events in the contemporary history of this technology. The media coverage that resulted from the opening of the first dedicated food irradiation plant in Florida and the Jack-in-the-Box hamburger poisonings combined to bring about a major shift in the public's perception of this technology. Food poisoning has started to enter the consciousness of the consumer as an issue of growing concern. Even commercial advertisements have begun to feature the problem of pathogens in the kitchen. Most importantly, the U.S. government has gone on the offensive regarding levels of pathogenic bacteria in raw food products. Although they may occur naturally, disease-causing microorganisms will no longer be tolerated with the same empathy as in the past.

On the technical side, considerable analytical work on the development of methods for detection of irradiated foods has taken place. Research to improve both the organoleptic and keeping qualities of irradiated foods continues to demonstrate the technology's value and practical potential.

Despite the interminable delays encountered in its full commercial utilization (as was the case with pasteurization), the prognosis for food irradiation appears to be very optimistic. The growing public concern for foodborne disease and the general change in attitude towards food irradiation have not been lost on the food industry – the sector ultimately responsible for bringing the products of this technology to the consumer. I have therefore added a chapter on key industry concerns to this edition. In this new chapter, irradiation is no longer viewed solely as a corrective or rectifying mechanism for stubborn industry problems, but rather, as a significant opportunity that must be placed in context with other major technological breakthroughs.

The interval between editions has also had its sorrowful moments. Dr. Ronald E. Engle, a pioneering spokesman on the use of food irradiation, passed away on May 11, 1994. President of the World Association of Veterinary Food Hygienists at the time of his death, ''Skip'' will be sadly missed by all his colleagues.

When my daughter, Heather, was two years old she contracted sal-monellosis as a result of eating an egg sandwich at a local snack bar. The other people who ate the same food probably suffered no more than some stomach cramps or diarrhea, which they most likely put down to over-eating or to stomach flu. Unfortunately, Heather had a bad cold at the time, and was in a weakened condition. Because of this particular set of circumstances, instead of simply causing some minor intestinal problems, the *Salmonella montevideo* bacteria that came with the sandwich became septic and entered her bloodstream. Once there, they took on the characteristics of their killer cousins, *Salmonella typhosa*, the bacteria responsible for typhoid fever. For almost two weeks, my daughter was administered antibiotics as she lay helplessly on an ice-water mattress to keep her fever down. When the ordeal was finally over, my daughter survived. I still give thanks to God, Pasteur and modern medicine. Heather's only crime was to eat an egg sandwich.

I have spent most of my professional career in the food industry. Although I was familiar with food irradiation and its advantages at the time of Heather's illness, irradiated foods and the benefits they offered were simply not available to consumers who might want them. That was 20 years ago.

Immediately after the disaster at the Chernobyl nuclear plant in the late spring of 1986, a number of colleagues approached me to say that this latest nuclear catastrophe would spell the end of food irradiation. Despite my protests that Chernobyl had nothing at all to do with food irradiation, I always received the same response, ''Yes, of course food irradiation has nothing to do with Chernobyl—but the public doesn't know it. What's even worse, the public has no idea of the ongoing critical problems of sanitation and hygiene in the food industry – so why should they worry about something new and controversial?''

What can one say? It is true. Even though food is one of the basic essentials of everyday life, it is not a subject for which most people have shown much interest. We may read with curiosity an article describing the sumptuous dinner held at the White House or the Royal Palace. We will watch some pompous gourmet describing which wine to consume with Trippa Romana on the television. But we really know very little about food and how it gets to our table. And most probably, neither does the television gourmet. The little we do know is generally trivial or distorted, and is often based upon the most superficial sources of information.

Food irradiation, along with pasteurization, canning, freezing and drying is simply a method of treating food in order to make it safer to eat and longer lasting. But, as is the case with most other food preservation methods, consumer knowledge of the subject is very limited. Despite the desire to inform more consumers on the topic of food irradiation, other priorities and obligations compelled the subject to remain dormant in the back of my mind for some years. In the interim, the public debate on food irradiation continued – a subject of great emotion, rhetoric and interest to those directly involved, but of little concern to most consumers.

In the fall of 1988, Britain's junior minister of health rather bluntly brought to the public's attention the high levels of *Salmonella* contamination in poultry and eggs. Hysteria followed, and sales of the implicated foods plummeted sharply. It has been estimated that industry losses may have reached £100 million or even higher. Although the ensuing public controversy forced the junior minister to resign, the high levels of bacterial contamination she warned of remained and, indeed, the number of diagnosed cases of food poisoning shot up dramatically. Consumers had finally been forced to take notice of a food supply that still left much to be desired. They also started to realize that every bad stomachache they experienced wasn't necessarily the 24-hour flu – they could very well have been poisoned!

I was convinced at the time, that a growing public knowledge of some of the more critical problems in the food supply would work in favor of irradiation processing for certain foods. Despite its frightening name, consumers would surely learn that it was one of the safest and most effective methods of food processing ever developed.

Later that year, the International Conference on the Acceptance, Control of and Trade in Irradiated Food was held in Geneva under the auspices of the Food and Agriculture Organization of the United Nations, the World Health Organization, the International Atomic Energy Agency, and the International Trade Center. Almost all the government delegations included representatives of national consumer associations, and consumer groups were additionally represented by certain non-governmental organizations. The consumer representatives were rather vocal, and a considerable amount of time was spent on their particular concerns with food irradiation. As disturbing as it was to see consumers who lacked an awareness of the subject, it was far more distressing to come up against those who were totally misinformed. Some of them were so thoroughly convinced of their position that no amount of information could shake their confidence in the utterly false and misguided beliefs they stubbornly clung to.

This attitude is neither new nor unique. Despite his monumental contributions to the prevention of disease and spoilage, Louis Pasteur was viciously criticized when he first demonstrated that the bacteria responsible for these conditions developed through reproduction rather than *spontaneous generation*. His only defense was to maintain that the results of his work had nothing to do with religion or philosophy or materialism, they were simply science. Unfortunately, the knowledge and science that he so tirelessly devoted himself to were not sufficiently developed at the time to prevent the death of his own daughter from typhoid fever [1].

Yet, over a century later, we have still not fully accepted the notion that scientific advance is not a subject of religion, philosophy or materialism. Science is principally the result of objective and controlled observation. Scientific fact cannot be swept away by a barrage of moral assertions. The very words *food irradiation* have made the technology a subject upon which people feel they can voice principled opinions, even before they have basic facts. The issue of food irradiation has spawned a vast cadre of opportunists who are more than happy to fill the vacuum of consumer knowledge with emotive misinformation rather than established and verifiable facts. This unfortunate circumstance has ensured that consumers remained ignorant of most of the critical problems in the food industry as well as information on their potential solutions.

The emotion of the conference debate, the distorted interpretation of scientific information and the stubborn retention of ideas long ago disproved, dashed any hopes I may have had that consumers would soon benefit from the knowledge gained through years of scientific work on food irradiation. The whole exercise left me with the inescapable feeling that we seem destined to run our lives by ignorance rather than knowledge–the result of which was needless suffering and inconvenience.

A few months after the Geneva conference, I decided to write this book.

Food irradiation and the various issues which surround it are not simply subjects. In trying to present them in easy, understandable language, there is always the temptation to oversimplify matters. Aside from a less precise understanding of the key issues, this can also result in a significant loss of perspective. On the other hand, excessive detail can be quite boring and, if improperly presented, can result in a similar loss of perspective.

In writing this book, I have attempted to cover both the technology and its associated issues in sufficient detail to allow for a fairly comprehensive understanding of the subject. I have also tried to relate the subject to matters of particular public concern. When all is said and done, however, this work is only a synopsis of the thousands and thousands of publications on the subject of food irradiation.

Where possible, I have given full details of the publications I have referred to. The list of References at the back is in the standard scientific format, with the minor exception that the names of the journals are fully spelled out rather than abbreviated, in order for readers to look them up more easily. I would encourage readers to do so if they have the opportunity. Whereas it may have only been possible to cover a topic with minimum detail in this work, the original references will, most likely, provide many more particulars of interest.

In one way or another, many people contribute to the writing of a book. This work on food irradiation is certainly no exception. Without going into any detailed explanations, I would therefore like to thank Fritz Diehl, Paisan Loaharanu, Fritz Käferstein, "Skip" Engel, Björn Sigurbjörnsson, Christine Bruhn and George Giddings–friends and professionals one and all. I would also like to express my appreciation to Jan Leemhorst, Sam Whitney, David Laurenzo and Jim Corrigan, all of whom belong to the honor roll of those courageous entrepreneurs who make changes to the world we live in. Finally, I would like to thank Miriam, Tracy, Heather and Carey–Les Girls–for their patience, understanding and support throughout this exercise.

The earth is round. There are some people who may say that the earth is flat [2] – but all the scientific tests and observations carried out thus far clearly indicate that it is round. Although people are free to believe or say the earth is flat, they are not allowed to prevent others from traveling around it, or otherwise taking advantage of the scientific fact that the earth is round.

Scientifically accepted tests indicate that irradiated foods are safe. In fact, when all the scientifically accepted evidence is considered, it can legitimately be concluded that food irradiation is one of the safest methods of food preservation ever developed. Again, there are some people who, in opposition to the scientists, say that food irradiation is not safe. However, unlike the flat earth folk, these people have managed to prevent most consumers from taking advantage of scientifically accepted evidence.

Why is this? Why have consumers not been given free access to such a beneficial technology when all the accepted scientific evidence has clearly demonstrated its value and safety?

There are many reasons for the delay in the introduction of food irradiation. This book will try to examine them all. Hopefully, by the end of this exploration, the reader will have a greater understanding of the technology, the issues, and the conflicts of interest that have come into play to prevent access to food irradiation. However, since this is not the first time that politics, self-interest and ignorance have combined to abuse the consumer, parallels will be drawn with another safe and beneficial technology that also took decades before public access was allowed – pasteurization.

Although food irradiation is the main subject of this book, *the central issue of this book is consumer choice—free and informed consumer choice*. In a world where most societies claim to have achieved the finest model of democracy, it is generally understood that we all have the basic right to choose. We choose the clothes we wear, the food we eat, the way we travel and the books we read. In a democracy, it is implicit that we also recognize and respect the rights of others to choose. We understand that choices are often simply value judgments. That is why Lucretius said over 2,000 years ago, "What is food to one, is to others bitter poison."

There are many reasons for making particular choices. We choose things because of sheer necessity, or because of the status they give us. We choose products because of their performance, or value, or safety. In making our decisions, most of us try to get as much information as we can in order to have some confidence in our choices. In particular, we want to be fully aware of any problems that might be associated with one choice or another. Easy access to the information we want is not always possible and we often rely on our elected representatives—the government—to make sure we don't make any poor or even dangerous choices, particularly where they may affect our health and safety.

In order to encourage rational and safe choices, noncommercial information is often provided to consumers. In the past few decades, as a result of pressure from consumer lobbying groups, far more information is available than ever before, especially on foods. It is presumed that this information is honest and unbiased. If not, our consumer rights have been violated and we complain to the retailers, manufacturers or even our elected representatives. Some even seek recourse in the public media to put things right. As a last resort, there are usually laws in place to protect us against fraudulent or misleading information and products. When the system works well, we end up with a wonderful result—a free and informed consumer choice. This free choice, aside from reflecting our increasing prosperity, has really become a symbol of democracy.

Not surprisingly, the system isn't perfect. There continue to be abuses. Some manufacturers give wrong or misleading information. This can be the result of inexcusable ignorance or the wish to take short-term advantage of fads or legal loopholes or, in the worst case, it is the result of an unconscionable cynicism towards an unsuspecting public. Manufacturers are not the only guilty parties. Some consumer advocacy groups are guilty of the very same faults. Being a consumer advocate doesn't release anyone from the responsibility of representing all the facts honestly. Therefore, dishonest or irresponsible consumer advocates must be regarded with the same disdain as their counterparts in the manufacturing industry. To confuse matters further, the press at times can slant an article, or choose to place undue emphasis on "sensational" issues and thus contribute to a loss of perspective for their readers. This apparently sells more newspapers, but it also results in misinformation rather than information, and is simply another example of promoting short-term self-interest to the detriment of the consuming public.

Even when full information is provided, there is no guarantee consumers will make wise decisions. Although there are clear health warnings on the packaging, many people still smoke cigarettes. In many countries people continue to risk the consequences of various pathogenic microorganisms by consuming raw milk, despite a full century of experience with the benefits of pasteurization. Old habits are hard to break even when people are fully aware of their potential harm. It can take generations to break old habits, but common sense eventually prevails; a free and informed choice is essential to foster and develop that sense.

Nowhere is this more apparent than in our choice of foods. In most countries there is a well-established system in place to make sure that food is safe and that consumers know as much about it as possible. There is legislation controlling the quality as well as the information we are given about the nutritional value of our foods. In order to confidently rely upon this system, we resort to the use of trained professionals—people who have the specific and lengthy education to make them qualified to do their jobs—nutritionists, food technologists, home economists, health inspectors, etc.

It is both sad and ironic that this high degree of specialization often results in the alienation of professionals from consumers, but it seems almost inevitable. The level of expertise needed to perform highly specific tasks leaves the professional little time to spend in daily contact with consumers. The result is that most scientists employed in the food industry are very poor at communicating with the public. They often resort to complex jargon, which may be second nature to them, but is a confusing mystery to most consumers. Scientists seldom relate to the fears and concerns of consumers simply because they have a technical understanding of the issues that consumers don't. It is an unfortunate circumstance, because this alienation from the consumer leads to a suspicion of both scientists and their technologies. Although this may be understandable, it is neither fair nor justified. Scientists are obviously no different from anyone else. They have parents, spouses and children to look after. They are also consumers. The difference is that they are informed consumers and are consequently more confident in their understanding of particular issues. The problem is that they are not particularly effective in conveying this confidence and understanding to consumers.

The gap between scientists and consumers leaves ample room for a host of misinformers to intercede. There is little doubt that a significant proportion of these people serve their clients out of a genuine sense of duty—but this does not justify the misinterpretation of facts and evidence, whether done intentionally or not. Social reformers, single-issue advocates and the ever-present "independent consultants" are all part of this following. Their livelihood is ensured through controversy, and their tools are the enhancement and exploitation of fear of the unknown. They are not a new phenomenon. Such "experts" have been around since time immemorial. Their specialty is the avoidance of accepted fact and its replacement with fearful supposition. They often refer to the simpler and better times of the past and warn of future impending doom. It all sounds so plausible, but in most cases it is based on neither fact nor experience. Adherence to their advice would lead us all back to a life expectancy of forty unhealthy years. Although some advocacy groups or newspapers may occasionally choose to use these untrained "experts," who sport backgrounds in totally unrelated disciplines, most governments cannot. Governments are responsible for the safety and well-being of their citizens, and cannot afford to take chances when it comes to the provision and interpretation of basic information. Of the many who would like to give advice and information on food in order to influence our choices, responsible governments can only rely on fully qualified specialists to make safety judgments and to provide accurate information. In practice, this does not always happen because of politics or self-interest, but the system to provide reliable information is generally in place in most governments.

It is no different from the system and approach most people follow in their personal lives. This is why we ask accountants to prepare our taxes, and plumbers to fix our water pipes, not the other way around. It is not always a foolproof approach, but generally speaking, highly specific tasks are best entrusted to highly qualified specialists.

Salmonella roulette.

It is a system that few of us object to. We may, of course, object to specific examples of questionable performance, such as misinformation, lack of information, or insufficiently strict or biased food laws, but we are generally happy to have a system that informs consumers about their prospective choices, and which does its best to protect the public from potential harm. The last thing most of us want to do is to gamble on the safety of the foods we eat.

Speaking of gambling — who hasn't seen some version of the old shell game? This is where a professional gambler puts a number of objects under some cups or shells, moves them around a table, interchanges them quickly, and finally asks you to choose the shell under which a particular object rests. Although it is considered to be a gamble, it isn't pure chance, because if the hand is not quicker than your eye, you may very well choose the right shell. It is a sort of free and semi-informed choice.

With the picture of the shell game still in mind, can you imagine going into a food store and innocently asking for a chicken? The butcher then dramatically presents you with three inverted pots. Under each pot is a chicken. He tells you that under one of the pots is a chicken contaminated with *Salmonella* bacteria. He gives you a fleeting glance at the contaminated chicken and then starts to move the pots around with ever-increasing velocity. Under two of the whirling pots are chickens that might end up as coq au vin, barbecued chicken or chicken Kiev. Under the other pot is a chicken that might end up as cramps, diarrhea or some other miserable and humiliating affliction. Which pot do you choose? Welcome to the game of Salmonella roulette!

However exaggerated the hypothetical game of Salmonella roulette may sound, in real life, your chances are not that good. In Salmonella roulette, the butcher kindly gave you the chance to see the *Salmonella*-laden chicken. In real life, you are not given that privilege. You have no way of knowing which is the contaminated chicken. In Salmonella roulette, only one chicken out of three was infected with *Salmonella*. In real life one, two or even all three of the chickens will be *Salmonella* positive. And, in real life, you will never find a game of Salmonella roulette where all the birds are free of *Salmonella*.

Thus, if you wanted to go into a food store and be presented with a free and informed choice of *Salmonella*-reduced or *Salmonella*-contaminated chicken, you would be hard-pressed to do it at this time. In real life, consumers are not given nearly the same chance as in the hypothetical game. But there are other games in town.

One could have a similar shell game with milk. Raw milk, which often harbors *Salmonella* bacteria, is placed under one shell, and pasteurized milk under the two others. Fortunately, in many countries where both types of milk are sold, they are clearly and properly labeled. You stand a much better chance at choosing the product you want and, if you choose, you can completely eliminate the risk of milkborne infection by ignoring the gamble and buying pasteurized milk. Which would you choose? That is up to you. Whatever your choice, you have been allowed to distinguish between each product, and you have been given options. You have been given the right to choose.

Why have consumers not been provided the same rights with chicken? Poultry is one of our most nutritious, versatile and economical food choices. Why have consumers not been afforded the same option of protection from foodborne diseases in chicken as they have with milk? And why have consumers not been given the right to choose this option? Why are we all forced to play roulette?

It is certainly not because there is no technology available that can accomplish this, because there is — food irradiation, technology that recognized experts agree is safe, effective and beneficial. Why then, have we not been given the option? Why is there no choice?

It is because there is a dreadful public confusion and misunderstanding about the process. As a result, the plumbers have messed up our taxes, the accountants have blocked our water pipes, the decision makers have sidestepped their responsibilities, and we have been denied a basic right – a free and informed choice.

CHAPTER 1

Food Irradiation

A prerequisite to appreciating all the issues involved in the food irradiation debate is an understanding of the technology and its history. But prior to launching into this subject it is important to examine the name of the technology since this itself has played an important role in the controversy.

There is no doubt that the name *food irradiation* makes most people immediately think of nuclear radiation. Many of us have seen the terrible pictures of Hiroshima and Nagasaki after they were hit with the first atomic bombs. Since that first episode, the threat of nuclear war has been our constant fear. Despite the end of the cold war and the warming of relationships between all the great military powers, the fear of a nuclear confrontation still plagues us. In fact, the dismantling of the former Soviet Union has resulted in a weakened central infrastructure to control the vast number of nuclear weapons still in existence. Recent reports of the attempted smuggling of weapons-grade uranium out of Russia and Eastern Europe are an inevitable result of this situation. As a consequence, the chance of some radical power getting hold of this technology and using it as a means of blackmail or mass destruction is still within the realm of possibility.

Aside from its obvious connection to the atomic bomb, nuclear radiation is also a fear associated with the nuclear power industry. Notwithstanding the benefits that have already accrued from the use of nuclear energy, the specters of Three Mile Island and Chernobyl continue to haunt us—and rightly so.

Many countries have chosen to be nuclear free in an attempt to avoid

1

any possible accidents from such power plants in the future. In making such decisions, the risks of the technology are normally weighed off against the benefits. It must be understood, however, that risks and benefits are not absolute values. Depending upon time, the technology in question, and the general circumstances of the world, the risks and benefits can change. Risks can be reduced through improved or more reliable technology. Benefits can increase when other alternatives become less attractive; for example the environmental buildup of acid rain due to high-sulphur coal burning. Unfortunately, self-interest and power politics often exert a far greater influence than they should in the final assessment of risks versus possible benefits.

Nuclear-powered electricity plants were designed to replace coal-powered plants because they were cheaper and caused less apparent pollution. Nevertheless, many countries feel their conventional sources of energy are more suitable, and currently prefer the risks associated with them, rather than with nuclear energy. Only time will tell if these judgments will change. There is no question in anyone's mind, however, that accidentally released radioactivity or uncontrolled nuclear radiation is a very frightening affair. Fortunately, it has nothing to do with irradiated foods. Unfortunately, it has very much to do with the erroneous perception of what food irradiation actually is.

In fact, the mistaken association of food irradiation with nuclear radiation is so great that food irradiation proponents resent the name and blame it for all the public's irrational fears of the process. They would love to change the name, not to conceal the technology, but to remove its frequent association and confusion with nuclear radiation.

On the other hand, opponents adore the term *food irradiation*. Since all the scientifically accepted evidence supports the safety of irradiated foods, association of the process with all the fears of nuclear radiation has become their most effective tool of negative influence. The classic question asked is ''Do you want to have your food nuked?'' The very name of the process is enough to frighten anyone who knows little about it. Jokes about glowing or radioactive foods are not funny for those who actually believe them. Thus, the term *food irradiation* itself, has been part of the problem.

If, however, there is sufficient knowledge of the subject to prevent confusion and misunderstanding, there should be no reason to change or avoid the name. Thus, in the optimistic hope of being able to convey a deep enough understanding of the process to eliminate any confusion with nuclear radiation, the term *food irradiation* will be used freely throughout this book.

(Because gamma rays ionize certain molecules during food irradiation, the process is quite correctly called *ionization* in France. The term *irradiation* is so intimately associated with nuclear radioactivity in the French language that its use, in relation to food processing, is grossly misleading. As a result, *ionization* has become the generally understood term equivalent to food irradiation in the French language. The same argument can be applied to other languages.)

What Is Irradiation?

Before reviewing the history and applications of food irradiation, it is first necessary to understand what the process actually is. Unfortunately, a certain amount of technical jargon is unavoidable in explaining the subject.

In order to understand the term *irradiation*, it is first necessary to understand the word *radiation*.

The *Van Nostrand's Scientific Encyclopedia* defines *radiation* as follows [3]:

Radiation

1. The emission and propagation of energy through space or through a material medium in the form of waves: for instance the emission of electromagnetic waves, or sound and elastic waves.

2. . . . The term radiation or radiant energy, when unqualified, usually refers to electromagnetic radiation; such radiation is commonly classified according to frequency, as radio frequency, microwave, infrared, visible (light), ultra-violet, x-rays, and γ (gamma)-rays.

In fact, the most common use of the term *radiation*, in both a scientist's as well as a layman's sense, refers to waves or rays in the electromagnetic spectrum. The electromagnetic spectrum is the full range of frequencies or wavelengths of electromagnetic radiation. This is most commonly pictured as a ruler along which the range of frequencies is spread. Many of the regions of the spectrum are familiar to most of us. Microwaves, infrared, ultraviolet and X-rays are all fairly common sources of energy we regularly encounter. The electromagnetic spectrum is diagrammed in Figure 1.1.

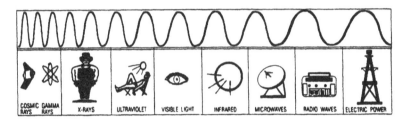

FIGURE 1.1 The electromagnetic spectrum.

Therefore, irradiation refers either to exposure to, or illumination by, rays or waves of all types. The *Webster Dictionary* describes *irradiate* as follows [4]:

> To illuminate or shed light upon; to cast splendor or brilliancy upon, to illuminate; to penetrate by radiation; to treat for healing by radiation, as that by X-rays or ultraviolet rays.

When you lie in the sun to get a tan, you are being irradiated. Depending upon the sensitivity of your skin, and the intensity of exposure, this form of irradiation can simply make you look good or it can make you very ill. Common sense and the use of protective suntan lotions have lessened the risks of skin disease associated with excess exposure to the sun's radiation, and this pastime is still enjoyed by millions of people. If the suntan lotions had to be (accurately) called *radiation protection cream* or *irradiation lotion*, it would definitely put people off. However, most would eventually return to the use of these lotions once they knew what they were, and understood the benefits of their use.

In fact, as will be discussed later, the term *irradiation* has been used in precisely this way in the past. Vitamin D is naturally formed by the action of sunlight upon our skin. In order to improve its vitamin D content, milk was exposed to ultraviolet light and was sold as irradiated milk [5]. This ultraviolet irradiation process is still in use in some parts of the world.

Visible light is generally considered to be the range of wavelengths in the electromagnetic spectrum which are least hazardous to humans. This is simply because light doesn't penetrate deeply past our protective skin, and consequently does not affect our sensitive internal organs. Microwaves, X-rays, gamma-rays, and cosmic rays, on the other hand, have greater penetrating power and can be dangerous. If one receives enough exposure directly from these rays, one will certainly be injured and possibly even die. That is why people would no more walk into a microwave oven than an irradiation chamber. That is also why strict standards of personnel control and exposure are applied to all forms of penetrating radiation.

Radiation of various wavelengths is employed to carry out tasks which could not normally be done otherwise. For instance, the convenience of microwave ovens allows us to heat up foods and beverages more rapidly than ever before. The radiation fears people originally had of these ovens have been overcome simply by time, familiarity and their routine presence on the market. People may differ in their opinions as to how well microwaves perform in the cooking of various foods. Some people love cakes made in the microwave, and others wouldn't touch them. However, no one can deny that these ovens can heat up foods with extraordinary speed. This can be very convenient after a long day at work, or when heating up various prepared dishes for company. Yet, there was a time when there was a genuine doubt if the public would overcome its fear of microwave ovens and the foods prepared in them. There are still people who distrust microwave ovens and don't use them. That is their right and their choice.

Ultraviolet radiation, on the other side of the electromagnetic spectrum, is used for many things. In the pharmaceutical industry, it is used to keep rooms sterile, and is employed in certain critical areas of the food industry to minimize microbial problems. Ultraviolet radiation is used to cure certain epoxy glues and resins. Degradable plastic packaging, designed to reduce environmental clutter, depends upon the ultraviolet rays that are present in sunlight to achieve its desired result.

Almost everyone is familiar with the radiation from X-rays, shown in Figure 1.2. Soon after their accidental discovery by the German physicist, Wilhelm Roentgen, in 1895, the penetrating power of X-rays was put to use by physicians to examine our bodies without surgical invasion. Damaged organs, broken bones and cancerous tissues can all be detected prior to treatment. The newer CAT (computer aided tomography) scanners have brought this beneficial technology to unparalleled levels of sophistication.

X-rays are also used to individually examine the thousands of parts that make up the jet engines that power today's airplanes. The unique ability of X-rays to detect stress cracks and other abnormalities has provided us with unparalleled reliability in these new engines. There still are occasional engine failures, but this modern technology, intelligently

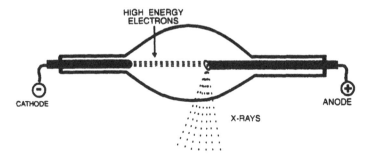

FIGURE 1.2 Generation of X-rays.

applied, has resulted in an overall level of performance that is truly astonishing. This level of reliability has also served to positively affect consumers' perception of the benefits versus the risks of air travel.

Radioactivity

In spite of the fact that the definition of the term *radiation* includes the entire electromagnetic spectrum, most people identify the word with *radioactivity*. There is also a popular belief that both radiation and radioactivity are relatively new phenomena. They are not.

Radiation and radioactivity are the result of the formation of our universe billions of years ago. Although they have been a normal part of our planet's existence since its birth, it has only been one century since we have become consciously aware of them.

It all started from the accidental discovery, in 1895, by French scientist Henri Becquerel, that natural uranium could affect photographic plates which were protected by lightproof paper. Within three years, Marie and Pierre Curie revealed the natural breakdown of uranium into the elements polonium (named after Poland, Marie's birthplace) and radium. This breakdown was accompanied by measurable *radiations*, which Marie called *radioactivity*.

From the very start of radiation research, the dangers of exposure to radioactivity were known. In fact, both Henri Becquerel and Marie Curie suffered from effects of direct exposure. Despite these hazards, research continued both into the peaceful and the military uses of this new discovery.

In reality, most of the work was directed at trying to reveal the physical structure of atoms. In order to grasp how radioactivity is formed, it is essential to visualize what atoms are like. The easiest way to picture atoms is to imagine a solar system. Just as the sun is the center of our solar system, so is the nucleus the center of the atom. The nucleus is extremely dense and is made up of closely packed heavy particles— protons and neutrons. Proton particles carry a positive electrical charge and neutrons, as their name suggests, are neutral— they don't carry any charge. In order to balance off this central positive charge in the nucleus, atoms have electrons which spin around the nucleus, in much the same way as our planets orbit around the sun. This results in atoms with a neutral electrical charge, as diagrammed in Figure 1.3.

It is the protons and electrons that determine which element the atom actually is. For instance, hydrogen is the lightest atom because it has only one proton in the nucleus and one electron spinning around in orbit. Since electrons contribute virtually nothing to the weight of the atom, both the atomic number and atomic weight are simply one.

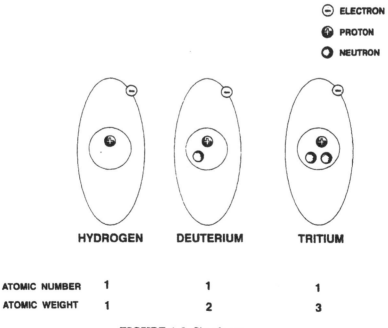

ATOMIC NUMBER	1	1	1
ATOMIC WEIGHT	1	2	3

FIGURE 1.3 Simple atoms.

But there can be natural variations of hydrogen atoms if more neutrons can be packed into the nucleus. For example, if an atom has a nucleus with one proton and one neutron, it still requires only one electron in orbit for the atom to remain neutral. While the atomic number remains one because of the one proton, the atomic weight will be two, because neutrons weigh as much as protons. This variation of the hydrogen atom is called *deuterium*. It is also called an isotope of hydrogen because it has the same atomic number, but a different atomic weight. Tritium is another natural isotope of hydrogen and contains the same single proton, along with two neutrons packed in the nucleus. The atomic number is still only one, but because of the two additional neutrons, the atomic weight is now three. All three isotopes are variations of the same element – hydrogen.

Uranium, on the other hand, is a much larger atom and contains 92 protons in the nucleus and 92 electrons spinning around in various orbits. The most common form of uranium also has 146 neutrons packed into the nucleus, so that the atomic number is 92, and the atomic weight is 238. Therefore, it is commonly called U^{238}. Other naturally occurring isotopes of uranium are U^{234} and U^{235}, which, as you can easily calculate, have 142 and 143 neutrons respectively. Employing modern nuclear technology, it is also possible to artificially produce certain isotopes such as U^{236}, U^{237}, U^{239} and U^{240}.

Although some natural isotopes are stable, many are not, and exhibit a tendency to change their form. As an example, the neutrons and protons in the nucleus of U^{238} barely hold together. They can't wait to disassociate and at some point, a particle made up of two protons and two neutrons breaks loose and U^{238} transforms itself into another element – thorium^{234}, a process diagrammed in Figure 1.4. This sudden change results in a release of energy, which is transmitted as radiation. The small particle which broke away, the two protons and two neutrons, is called alpha (α) radiation.

Thorium^{234}, which now contains 90 protons and 144 neutrons, also undergoes a change because it, too, is unstable. Through a different and more complicated mechanism, it transforms one of its neutrons into a proton, thus changing the nucleus, so that it contains 91 protons and 143 neutrons. At the same time, the atom discharges an electron in a process called beta (β) radiation.

The new atom is called protactinium^{234} (see Figure 1.5), which exhibits the same unstable tendencies as the rest of the family and continues

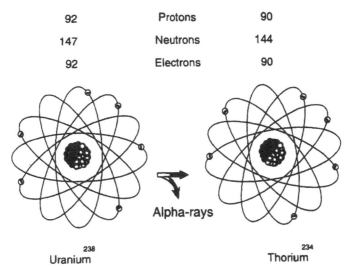

92	Protons	90
147	Neutrons	144
92	Electrons	90

Alpha-rays

Uranium ^{238} Thorium ^{234}

FIGURE 1.4 Uranium and thorium atoms.

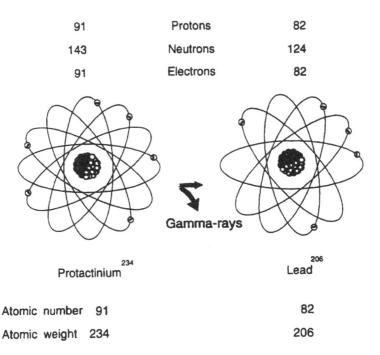

91	Protons	82
143	Neutrons	124
91	Electrons	82

Gamma-rays

Protactinium ^{234} Lead ^{206}

Atomic number 91   82

Atomic weight 234   206

FIGURE 1.5 Protactinium and lead atoms.

to transform. This series of natural transformations goes on until a stable element is finally formed. This happens to be lead^{206}, which has 82 protons and 124 neutrons.

During this series of events, there is occasionally a time when an isotope is so unstable that, in addition to α- or β-particles it gives off a burst of pure, non-particle energy called gamma (γ) radiation.

This whole transformation process, accompanied by the various atomic radiations is called *radioactivity*. The radiation produced as a result of this mechanism is the nuclear radiation we commonly talk about. The original names of the various rays were simply based upon their direction of travel in a magnetic field. Alpha (α)-rays always headed towards the north pole of a magnet, beta (β)-rays towards the south pole, and gamma (γ)-rays went straight down the middle (see Figure 1.6).

The α-, β- and γ-rays which are emitted have very different energy levels and differ in their ability to penetrate materials. If, for the purposes of comparison, we give a value of 1 for the penetrating power of α-rays, then β-rays would have a value of 100, and γ-rays would have a value of 10,000. While α-rays will hardly penetrate the surface layers of skin, β-rays (depending upon whether they result from radioactivity or electron accelerators) can penetrate anything from a thin sheet of paper to a finger's thickness of living tissue. Gamma-rays, on the other hand, are so energetic they can only be stopped by a heavy sheet of lead or several feet of water or concrete (Figure 1.7).

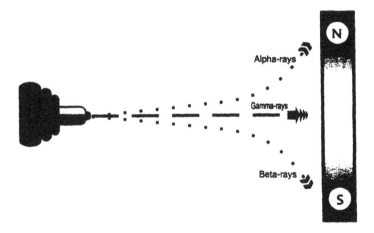

FIGURE 1.6 Radiation in a magnetic field.

ATOM

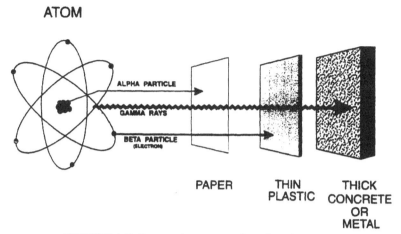

PAPER THIN PLASTIC THICK CONCRETE OR METAL

FIGURE 1.7 Penetrating power of α-, β- and γ-rays.

As previously mentioned, radioactivity and radiation occur normally in nature. In fact, the greatest exposure we routinely have to radioactivity is from natural sources. Quite a large fraction of this natural radiation comes from outer space, and is called cosmic radiation. As a result, people living in high mountain ranges such as the Himalayas or Andes will receive almost ten times as much exposure as people living at sea level. For the same reason, flying in airplanes also increases exposure, and aircraft personnel receive much higher exposures than other people. There are also inhabited areas of the earth where unusually high levels of radioactivity occur, such as those near deposits of natural, thorium-rich sands. The effects of long-term exposure to these higher levels of radioactivity is generally not known. Such levels may be harmful but are possibly masked by other factors—witness, for example, the legendary long life span for people living at high elevations. What is known without the slightest doubt is that exposure to excessive levels of high-energy radiation is definitely very dangerous.

How does the irradiation process work?

Since this book is concerned with food irradiation, we will now focus on the type of energy that is most commonly used in the process, namely gamma (γ) radiation. It should also be noted that X-rays and the more recent machine-generated electron beams can also be used to irradiate foods because they act in a very similar manner.

Gamma-rays, like X-rays and high-energy electron beams, are all called *ionizing* radiation because they are capable of knocking electrons out of their normal orbits in atoms or molecules. This action results in an atom or molecule that is no longer electrically neutral, and which goes looking for another electron to balance itself out again. There are other ways nature has provided for atoms or molecules to lose or even gain electrons in the course of their normal reactions. Whenever atoms or molecules are in this electrically charged state (since an electron can be added or lost), they are called *ions* or *free radicals* and are thus said to have been *ionized*. Free radicals are an intermediate stage of most common reactions that routinely occur in nature.

It must be clear that all changes which occur to matter involve reactions of some type. When you fry an egg, or toast bread, or pasteurize milk, or digest your food, you are changing the nature of matter. You are thus producing free radicals in the process. In fact, free radicals are routinely formed during normal metabolism [6] because free-radical reactions are one of the basic biochemical reactions of the body.

In this free radical state, atoms are very reactive and combine with other free radicals or other substances. Ionizing radiation is therefore another means of ionizing atoms or molecules prior to carrying out reactions. Heat or light can do the very same thing. Gamma (γ) radiation is a particularly effective method of achieving this since its penetrating power allows it to ionize atoms or molecules uniformly throughout a material. Although processes such as cooking also produce free radical transformations, heat does not have the same uniform penetrating power, and by the time you achieve the desired thermal effect in the middle of the product, the outside may be overcooked. Sometimes, you can take advantage of this effect, for example when you cook a rare roast beef or a 3-minute egg. The outermost portions will be cooked, but the center will be raw. At other times, the poor penetrating power of heat can be a nuisance. Unless soup is well stirred, the portion touching the hot bottom of the saucepan can burn before the rest gets hot. The poor penetrating power and distribution of heat can even be dangerous, as in the case of meats, which can be a source of bacterial or parasitic infection, if not fully cooked throughout.

Ionization and free radical formation are a normal part of all reactions that occur in nature, including the transformations that result during the processing of foods. By properly adjusting a process to carry out a particular transformation, it is possible to achieve a specific, desirable effect. The case of rare roast beef given above is a typical example of this. In the use of food irradiation, a number of changes may be desirable depending on the food product in question. For instance, it may be necessary to reduce the level of infectious bacteria that often occur in certain meat products. The products in question would therefore be subjected only to that amount of ionizing radiation sufficient to ensure that the pathogenic (disease-causing) bacteria are either dead or incapable of causing infection. The result is a meat product which is safe from the bacteria, but has not been cooked. Of course, one could get the same effect by fully precooking the meat, but few consumers would want to go into a supermarket and be presented exclusively with cooked products.

In another instance, certain vegetables, such as potatoes, have a tendency to sprout during storage because of active cell division. In the spontaneous process of getting the potato ready to germinate and grow into a new plant, these cell divisions cause dramatic changes in structure and chemical makeup. Starch, which is the potato's storage carbohydrate, is broken down into glucose, which the new plant will use for energy. The sprouting process makes the potatoes too soft and sweet for eating. In this case, food irradiation is used to ionize the responsible macromolecules, so that the cells are no longer able to cause sprouting. Again, the macromolecules can be inactivated by precooking, but it is difficult to imagine consumers finding boiled potatoes acceptable at the retail level. There are, in fact, other alternatives to food irradiation currently on the market, but, as shall be seen later, they may offer less benefit and greater risk to the consumer.

In developing the technology to make useful modifications to foods, it is essential to minimize any undesirable effects or changes which can result from the process. For example, when UHT (ultra high temperature) pasteurized milk was first introduced, it had a very obvious burnt or scorched taste. It took some time before this undesirable taste was minimized. When quick frozen vegetables were first produced and sold, it was not fully understood that some natural enzymes continued to be active, even below freezing temperatures. Consequently, even though they were stored frozen, significant deterioration in taste and texture occurred in these products. This enzyme effect was eventually overcome through a quick dip in very hot water (blanching) before freezing, which destroyed the enzyme without fully cooking the vegetables. (The same blanching should be done when freezing vegetables at home.)

In order to achieve the desired effects in the process of food irradiation, the products are exposed to ionizing radiation in a highly controlled manner. The most common type of irradiator uses the isotope cobalt[60] as the source of radiation. This is a man-made material which first starts off with highly purified, natural nonradioactive cobalt[59]. It is then tightly compressed into small cylindrical pellets, which are carefully fit into stainless steel tubes not much larger than pencils. These pencil tubes are placed in a nuclear reactor where they are constantly bombarded by neutrons for about one year. This process results in pellets of highly purified cobalt[60] which produce a controlled emission of γ-rays during their transformation down to a stable state of nickel[60]. The cobalt[60] used in food irradiation is the same as that used in medical irradiators, and is the product of a very sophisticated engineering, manufacturing and quality control process. It is definitely not a nuclear waste product as some people have foolishly claimed.

These high-energy γ-rays are then used in a facility which is specifically designed to irradiate products. The design of an irradiator is fairly straightforward. The irradiation source (Co^{60}) is located in the irradiation chamber and is stored in a protective environment when not in use. When required, the source is raised out of its shielding so that it can treat the products in question. The protective environment can be a pool of water which completely absorbs the energy, or it may even be the thick protective lead casing in which the source was originally shipped. The large irradiation chamber itself is constructed with thick concrete walls in order to absorb all γ-rays which are not absorbed by the products (see Figure 1.8).

Irradiation sources can also be made with Cs^{137}, an isotope of cesium. This material is obtained by reprocessing and extracting spent-fuel rods from nuclear reactors. The use of this material has brought on much criticism from antinuclear opponents who claim that food irradiation was simply invented to get rid of nuclear waste. While it may appear cheaper to extract Cs^{137} from spent-fuel rods for use as a source of γ-rays, the practical capacity to do so is very limited and will probably remain that way for some time in the future. In addition, the very act of reprocessing nuclear waste is so politically controversial that the future availability and common use of Cs^{137} as a food irradiation source is very unlikely.

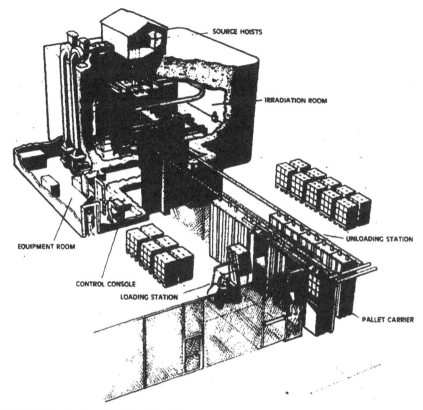

FIGURE 1.8 A typical food irradiation plant. (Courtesy of Gammaster B.V. and Nordion International, Inc.)

There are other practical reasons for the preferred use of Co^{60} as an irradiation source, including a greater degree of overall efficiency, better γ-ray penetration and greater environmental safety, due to its complete insolubility in water. As a result, Cs^{137} irradiators represent an extremely small proportion of today's irradiators, and are not used at all for commercial food irradiation.

A third method of irradiating food involves the use of electrical machine sources of energy. X-rays were among the first sources of radiation to be tested on foods. More modern techniques use machine sources called accelerators, which produce extremely high voltages of electron beams to shower the food to be processed. An obvious advantage of such systems is that they can be switched on and off like a light bulb, and are in no way related to the nuclear industry. A disad-

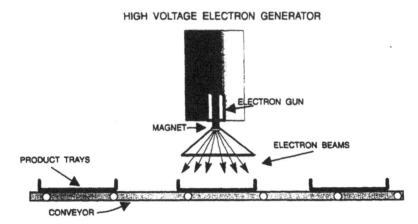

FIGURE 1.9 Electron beam irradiation.

vantage of these systems is that the high-energy electron beams have a limited penetrating power compared to γ-rays from a Co⁶⁰ source. Typically the penetrating power of electron beams is suitable only for materials with a thickness of no more than 2−4 inches (5−10 cm). For certain uses however, machine sources have proven to be very practical. In France, commercial quantities of deboned chicken meat are irradiated using this method. In the Black Sea port of Odessa for example, machine sources have been used on a large scale (200 metric tons of wheat per hour) for over a decade, in order to carry out insect disinfestation of imported grain.

The choice of source and irradiator design is therefore dependent upon the products to be irradiated, the cost of equipment, the operating expenses, and the sociopolitical environment in which the unit will function. Regardless of the source, however, the function of food irradiation is to protect or improve the quality of the food product, and it does so through uniform ionization.

Now that there is a basic idea of how the process works, the most important principle to understand is that food irradiation does not make foods radioactive.

How Is It Similar to Other Processes?

Foods are made up of four basic materials: water, carbohydrate (starch and sugar), lipid (fat and oil), and protein. There are also other constituents such as vitamins, minerals and various minor or trace elements. Food irradiation, like all other processes, affects these materials by causing changes through ionization. The reactive ions or free radicals formed during food irradiation combine with other ions to achieve a more stable state. This process is called *radiolysis* and the products formed as a result are called *radiolytic products*.

Depending on the type of food irradiated, the radiolytic products could be formed immediately, or they could form slowly over an extended period of time. This is not difficult to understand. When any free radical is first formed, it seeks a partner to combine with. If it finds a suitable one, a stable product results. But it has to be mobile in order to go looking around in the first place. Since water is a fluid, materials which are dissolved in water, or which have high moisture contents, will provide considerable mobility for free radicals. Stable end products will thus be formed quickly. Very dry products, on the other hand, do not allow the same degree of mobility, and the length of time required for free radicals to form stable products is much longer.

There are a great variety of radiolytic products that can be formed as a result of irradiation, depending upon the food being treated, and the conditions employed during processing. However, very few of these radiolytic products are unique. Most radiolytic products formed as a result of food irradiation are identical to the ionization products formed by all the other traditional types of food processing. The simple act of frying an egg or making toast in the morning will form the same free radicals and radiolytic products at similar, or higher levels.

It is very important to understand that, although they were formed through the use of heat, these latter ionization products are identical to the radiolytic products formed during irradiation. Because they were formed during a thermal process, they can be called *thermolytic products*. They are all *ionization products* and the method employed to form them has no effect on their actual composition. They are the same products. (Water can be boiled with gas burners, electrical heaters, infrared lamps, or microwave ovens. Regardless of the source of energy used, the end result is boiled water.)

The ionization products which result from irradiation can also be formed by cooking, steaming, roasting, pasteurizing, freezing, grinding and other forms of food preparation. Free radicals and ionization products also occur during the natural ripening of fruits and vegetables, the sprouting of beans and the coagulation of yogurt. You produce free radicals and radiolytic products whenever you use a pressure cooker, a steamer, a macrobiotic dryer or a wok. Free radicals are produced in almost every type of transformation.

All the reliable scientific evidence indicates that free radicals and radiolytic products formed during food irradiation and most other common food processes do not pose problems to us. However, because the terms *free radicals* and *radiolytic products* are not commonly known, they can easily be taken out of context and made to sound frightening. One can also get the impression that they are uniquely the result of food irradiation. This is simply not so. These products are as much a part of nature as organically grown garlic is. The cereal or toast one eats for breakfast every day can have even greater levels of the same ionization products that irradiated foods may have. In fact, free-radical formations are one of the very basic biochemical reactions carried out by the body and are normal components of everyday metabolism.

Yet, throughout the public debate on food irradiation, the common occurrence of free radicals and ionization products in all foods has often been deliberately ignored. Whenever this has been the case, consumers have been left with the impression that these "radicals" only result from food irradiation. As a consequence, consumers are being misinformed, rather than informed. It should be clear that humans have routinely consumed free radicals and radiolytic products from the dawn of our existence. In fact, the only way to avoid them is to stop eating altogether.

Changes that occur to proteins, lipids, vitamins and carbohydrates as a result of irradiation are the same as, or less than, those changes caused by cooking, canning, pickling, freezing and drying. If you wish to preserve food, you must prevent it from spoiling. The method one chooses will naturally depend upon practical, economic, cultural and nutritional considerations. The wide range of alternatives currently available to us reflects both the ongoing and the historical attention we have paid to the subject. But the common understanding that foods must be safely and practically preserved has never before been a subject of debate.

There is no doubt that, given the choice, most people would prefer to eat fresh, ripe, natural foods right from the source, all the time. Since leaving the Garden of Eden, however, this has not been a practical possibility for most of us. That is why we have always devised ways of preserving our food in order to survive. If the method chosen results in a product we like, it will be a product of preference. If not, we will reject it. The famous 100-year-old preserved eggs of China, which have a soft black yolk and a gelled green white, are considered a delicacy by millions. In other countries, they are not a product of preference. The choice of preservation methods varies from time to time and country to country, but the necessity to preserve food has always been, and will continue to be, a routine of the human condition.

In What Way Does Irradiation Affect Foods?

Once more, the operational term is ionization. Large and complicated molecules or macromolecules, such as nucleic acids are directly or indirectly responsible for significant spoilage. Because these macromolecules are so large, the chances of their being hit and ionized by γ-ray energy are high. For these macromolecules to carry out their highly specific functions, their precise structure and composition must be kept absolutely intact. Ionization slightly modifies or breaks their structure and thus prevents these molecules from functioning normally.

When potatoes are irradiated, the macromolecules responsible for initiating sprouting are ionized by the γ-rays. The very complexity of these large molecules, which makes them so active, also makes them so sensitive. One slight change to their structure, and they can no longer perform properly. Therefore, very low exposures of γ-radiation are necessary to inactivate the macromolecules responsible for sprouting. The very same effect could also be carried out through the use of heat, as previously mentioned, but in order to inactivate all the cells responsible for sprouting, the whole potato would probably have to be cooked.

Among the most important uses of irradiation is the destruction or reduction of pathogenic bacteria in foods. The life and reproduction of these microorganisms is dependent upon their nucleic acids, DNA (deoxyribonucleic acid) and RNA (ribonucleic acid). Because they are both very large and complicated macromolecules, they are very sensitive to ionization. As a result, relatively small exposure to γ-radiation is required to dramatically reduce the pathogenic potential of these bacteria in foods.

Much the same can be said about spoilage or disease-causing yeasts, molds, protozoa (e.g., dysentery ameba) and even viruses which can be transmitted through food. The ability of these organisms to reproduce is dependent upon their DNA and RNA. The very slight changes that result from ionization are sufficient to prevent growth and reproduction.

Certain parasites go through a stage in their life cycles where they infect the tissues of food animals we eat. Some of these parasites can then reinfect people. When foods containing these parasites are irradiated, the parasites are either killed outright or rendered incapable of further growth and reinfection.

The current treatment given many foods in order to eradicate insects is not sufficient to kill or eliminate the insect eggs. Consumers who have found tiny beetles in their flour weeks or months after it was purchased, know what this is. Many consumers think that the clean fruit which they purchased and left out a few days ago has attracted little flies. In fact, in most instances the fruit did not attract the flies, the fruit was the source of the flies, which originally came in as fly eggs.

Again, irradiation is employed here to kill the insects and eggs, or prevent their development, while leaving the fruit as close to its natural state as possible. There are other methods to do this, but they are not considered as safe or effective in carrying out this task. As in previous cases, it is the ionization of the critical macromolecules that does the job.

Radiation Units

The amount of irradiation energy that a food absorbs is measured in units called *Grays* (Gy). The unit refers to the amount of energy that 1 kilogram of the product receives from the ionizing radiation. One Gy is equivalent to one *joule* (a unit of energy) per kilogram. The practical working range for food irradiation is generally from 50 Gy to as high as 10,000 Gy, depending upon the food in question and the effect desired. It is difficult to relate this level of ionizing energy to other common forms of energy. If we relate it to heat, the upper level of 10,000 Gy, or 10 kiloGray (10 kGy) is equal to the amount of energy required to raise the temperature of water by 2.4 °C. But this comparison doesn't tell the consumer very much, because the energy forms are so different. The reason this relatively small amount of energy is so effective is that it can inactivate sensitive targets (e.g., bacterial or parasital DNA) throughout the food that are the cause of spoilage and disease.

As previously mentioned, when food is cooked, heat penetrates it very slowly and unevenly. The temperature at the center of the food is often far lower than on the surface. A good deal of the energy is wasted simply by bringing the product up to the temperature at which cooking will take place. In addition, as the surface dries out, the decreased moisture level can make heat penetration even more difficult. On the other hand, irradiation penetrates a food completely and evenly so that a small amount of energy can accomplish the required task without the need for slow, thermal penetration.

It is also difficult to get an accurate perspective on the numbers involved in food irradiation units. The use of figures such as 10,000 Gy sounds high. When one refers back to the original units used, Rads (for radiation absorbed dose), the figure appears even higher, since 1 Gy = 100 Rads. As a result, the large numbers themselves can easily confound the issue. In fact, many anti-food irradiation advocates have used these large numbers to try and frighten consumers. It is rather unfair because it misleads and misinforms those who are unfamiliar with the terminology.

The numbers game can always be a source of confusion. Take an apple, for example. When one eats a small apple (150 g), it contains about 90 kilocalories (which are the regular calories we normally refer to). However, this is equivalent to 90,000 small calories, or 21,500 joules or even 215,000,000,000 ergs (a basic unit of energy). But it is still only a small apple! If you know a small apple will not make you fat, why should you worry about eating 215,000,000,000 ergs worth? In food irradiation, as in all other energy considerations, large numbers in themselves have little significance. It is what the numbers mean regarding the effect of absorbed dose that is important. The dose level recommended for use must do the job required, and the finished product must be safe. That is what counts.

From a practical point of view, there are three general application and dose categories that are referred to when foods are treated with ionizing radiation [7]:

(1) Low dose—up to ≈ 1 kGy
   - sprout inhibition
   - delay of ripening
   - insect disinfestation

(2) Medium dose—1 to 10 kGy
   - reduction of spoilage microorganisms
   - reduction of non-spore-forming pathogens
   - delay of ripening

(3) High dose—10 to 50 kGy
   - reduction of microorganisms to the point of sterility

Depending upon the specific goal of the treatment, trade-type names
have been given to these general ranges that relate more to the desired
function than the actual dose level per se. They are defined as follows:

Radurization—treatment of foods with a dose of ionizing radiation
sufficient to improve shelf life by reducing substantial quantities of
spoilage microorganisms

Radicidation—treatment of foods with a dose of ionizing radiation
sufficient to reduce the level of non-spore-forming pathogens, including
parasites, to an undetectable level

Radappertization—treatment of foods with a dose of radiation sufficient
to reduce the level of microorganism to the point of sterility

Although this profusion of terminology is somewhat perplexing, it is
merely part of the jargon that has evolved over the years to categorize
food irradiation according to the type of treatment supplied. They may
come into more popular use if and when products treated with ionizing
radiation become available to the consumer, much in the same way as
the term *pasteurization* has.

A Short History of Food Irradiation

One of the best reviews made of the early history of food irradiation
was written by E. S. Josephson in the *Journal of Food Safety* in 1983 [8].
A less extensive, but no less interesting, review appears in the recently
published book *Safety of Irradiated Foods* by J. F. Diehl [9]. From these
references it is quite clear that the idea of using irradiation immediately
followed the discovery of radioactivity in 1895 by Henri Becquerel. In
fact, in the very same year that Becquerel published his work, the
suggestion to use ionizing radiation to destroy microorganisms in food
was published in a German medical journal [10]. Within a few years,
patents describing the use of ionizing radiation to destroy bacteria in food
were issued in both the United States and Britain [11]. As Diehl notes,
the authors of the 1905 British patent *wanted to bring about an improvement in the condition of foodstuffs* and their keeping quality. They felt
that this technique was an advantage because the improvements could be
made without using any chemical additives—a concept which is just as
valid today. The technology could not be commercially considered,
however, because the irradiation sources used (radium) were not easily
available at the time.

The next applications were described in 1916 by a Mr. G. A. Runner who used X-rays to kill the insects, eggs and larvae in tobacco leaves in order to improve the quality of cigars [12] and later in 1921 by B. Schwartz who employed X-rays to eliminate the *Trichinosis* parasites found in pork [13].

Other studies and patents slowly followed. The major limitations were the cost and availability of practical ionization sources. Although X-rays proved to be effective in preserving ground beef, they were simply too expensive to be feasible. All other potential ionization sources were likewise too costly to be considered for practical, commercial use.

This situation continued until after World War II and the ensuing Atomic Age. Access to the use of isotopes as experimental ionization sources was greatly increased when spent-fuel rods from nuclear reactors became available. Suddenly, food irradiation achieved the potential of being a commercially feasible process. Considering the situation, the historical timing for the easy availability of low-cost ionization sources could not have been much different, but neither could it have been much worse. Even though the food irradiation process was developed 50 years earlier, it suddenly became inextricably linked with the atomic bomb and nuclear radiation in the minds of all who were not fully familiar with the technology. Josephson described it skillfully in his article in the *Journal of Food Safety*:

> The midwife attending the birth of food irradiation was the development of nuclear fission and its military use at Hiroshima and Nagasaki. This stigma has attached itself to food irradiation's origins and has dogged its progress in the United States and abroad ever since. It is likely that if food irradiation had been spawned as an outgrowth of medical applications of nuclear energy, the public today would be enjoying the benefits of this new method for preserving food. [8]

With the availability of commercially feasible ionization sources, research into the safety and applications of food irradiation began in earnest. The U.S. Atomic Energy Commission and the U.S. Army commenced a fully coordinated research program on food irradiation using spent-fuel rods from nuclear reactors as their first ionization sources. (The U.S. Army, along with the military establishments of many countries, has had a traditional interest in food research, and has been involved in the development or improvement of many of today's common technologies, such as canning.) For practical reasons, the spent-fuel rods were eventually replaced with Co^{60} and Cs^{137} ionization sources.

Because of cost, functionality, and environmental characteristics, however, Co^{60} and more recently, machine sources, have become the most common means of irradiating foods.

The initial research on food irradiation revealed both the problems and advantages of the technology. As an example, excessive ionization of meat resulted in poor taste, smell and texture. Research revealed that it was just as easy to over-irradiate foods as it was to overcook them. Certain meats developed a poor odor when irradiation treatment was too severe. Just as research had shown that frozen vegetables retained better texture and flavor when they were given a prior blanching, so it was shown that irradiated meat flavor was best when treatment was limited to the minimum levels needed to accomplish the required task. Research indicated which foods responded well to food irradiation and which did not.

Because of the constant public association of the process with nuclear radiation, a research priority was naturally placed upon determining the health and safety of irradiated foods. As a result, the safety of this technology was scientifically studied more than any other food process.

During the past two decades, the Food and Agriculture Organization of the United Nations (FAO), the International Atomic Energy Agency (IAEA), and The World Health Organization (WHO) have become intimately involved with the issue of food irradiation. This is understandable, since several aspects of food irradiation technology fall within their operating mandates. Among the principal activities of the IAEA is the promotion and encouragement of the peaceful uses of nuclear energy. The FAO has, among its many global responsibilities, the task of ensuring a worldwide reduction of food spoilage or postharvest losses, as well as the overall improvement of food quality, safety and nutrition. The WHO is particularly concerned because of its global role in public health improvement through the reduction of foodborne diseases.

Since the process holds the promise of significantly improving the quality of our food supply, these organizations initiated an international effort to investigate the safety and wholesomeness of irradiated foods. In 1970, the International Project in the Field of Food Irradiation, sponsored by FAO and IAEA, with WHO in an advisory role, was established in Karlsruhe, Germany. The resources of over twenty supporting countries were pooled to carry out chemical analyses and animal feeding studies on a wide range of irradiated foods such as wheat, rice, meat, fish, fruit and spices.

At the conclusion of the studies, the FAO/IAEA/WHO Joint Expert Committee on Irradiated Foods assessed the data and in 1980 decisively stated that:

> The irradiation of any food commodity up to an overall average dose of 10 kGy presents no toxicological hazard; hence toxicological testing of foods so treated is no longer required. [14]

This is a rather extraordinary statement which should be taken in context. Because food irradiation had been studied more than any other food process, and all the products found to be safe, there was no reason to consider them any different from other foods. All foods fall under the scrutiny of health and safety inspection, but there was no justifiable reason to single out irradiated foods as opposed to foods processed in other, conventional ways. In fact, many of the food processes we have used for thousands of years would never come close to the safety standards which have been established by irradiated foods.

According to Diehl [9], the Federal Research Center for Nutrition at Karlsruhe, Germany, now contains about 9,000 documents on all aspects of food irradiation. It is perhaps one of the most detailed collections of information on a single subject kept anywhere. It is currently in the process of being placed in a data base format for easier access to scientists and all others interested in the topic.

All this evidence, gathered from almost a century of scientific and technical research in the field, leads to the unavoidable conclusion that food irradiation is a safe, practical and beneficial process.

Why, then, are irradiated foods not available? What is the problem?

Perhaps the answer to this question can be found by looking at the history of another, rather famous, food process which we take for granted today, but was the subject of limitless emotional debate and heated rhetoric in the past. It was a process that promised tremendous advantages to consumers. It allowed food to be kept longer without going bad, and it prevented the spread of many foodborne diseases. It was a practical solution to a number of real and undeniable difficulties in the food industry. Yet, detractors of this technology were able to effectively delay its introduction at the cost of thousands of lives and needless misery to millions of others. In some countries, the technology was delayed by almost a century. In other countries, the technology is still not fully available. Let us look at pasteurization.

CHAPTER 2

Pasteurization

The Early Work

The idea of preserving foods through the reduction or destruction of the microorganisms they harbor is centuries old. In the eighteenth century, boiling was used to preserve meat extracts (1765 – Spallanzani) and vinegar (1782 – Scheele) [15]. Canning, one of the greatest advances in food technology, resulted from the work of a French confectioner by the name of Nicholas Appert in 1804. Napoleon offered a prize for improved methods of preserving foods for the French army. Appert found that by heating food in a metal container, and sealing it off from air, the food could be kept in an edible condition for a very long time. Appert submitted his technology and won the prize in 1809. He deeply believed in his work and promptly invested the prize money in the world's first cannery.

Although there was the knowledge that these methods worked well to preserve foods, the precise reasons as to how or why they worked were not understood.

A similar situation existed in the field of medicine at the time. Although it was possible to control the symptoms of certain diseases, physicians did not know their precise causes. Microorganisms and their effects were virtually unknown.

The work of Pasteur (Figure 2.1) changed all that. Following in the direction set by the Italians, Spallanzani and Fabroni, he felt there was a close link between diseases and the fermentation that so often accompanies food spoilage. In his push "to arrive at the knowledge of the

27

causes of putrid and contagious diseases'' [1], he carried out in-depth studies on the causes of spoilage in wine. These studies eventually led to the investigation of disease in silkworms, and ultimately to his researches on animal and human diseases. Pasteur's effort revolutionized our world, and was probably the single most important scientific work providing us with a means of controlling the effect that microorganisms have upon us.

From 1860–1864, his research clearly demonstrated that spoilage in wine was caused by simple microorganisms (Figures 2.2 and 2.3). In order to handle the problem, he recommended a level of heat treatment that was sufficient to inactivate the spoilage microorganisms (50–60°C), but not great enough to destroy the quality or character of the products processed [16]. This was, in fact, the key to his process. He could easily have boiled the wine in order to kill off all the microorganisms, but that would have destroyed the wine's taste as well. He

LOUIS PASTEUR

FIGURE 2.1 Louis Pasteur [1].

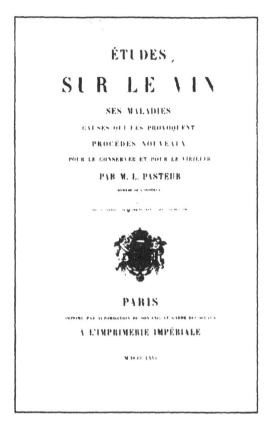

FIGURE 2.2 Front page of *Études*, Pasteur's original book on wine [16].

therefore determined the minimum processing required to do the job without impairing the product's overall acceptability. The technique turned out to be very successful and he soon applied it to beer as well.

Thus, for the first time, a method became available to prevent the spoilage that had always been accepted as an unavoidable fact of life. Consumers no longer had to tolerate wine or beer that spoiled prematurely. Not only was the shelf life of a product extended, but the very basis of its untimely spoilage was finally understood. Even though the commercial significance of this technology was the primary consideration at the time of its development, its greatest impact was to be ultimately felt in the area of public health. The drinking of wine or beer may not have been generally recommended for health, but the consumption of milk by children was, and it was in the dairy industry where

P. Lacenrbeurr, ad nat. del.   Imprimerie Impériale
                        Anali

FIGURE 2.3 Diagram from Pasteur's *Études Sur Le Vin* [16].

Pasteur's methods were to ultimately achieve their greatest success and
impact. His technique was a genuine breakthrough that saved the lives
of thousands and changed the course of our eating habits forever.
Curiously, there are no records of Pasteur himself employing the new
method on milk. However, when the same process was eventually
applied to milk, it was aptly named *Pasteurization* in recognition of his
enormous contribution to science and to our everyday lives.

Commercial Pasteurization

The pasteurization of milk did not become a commercial reality for
many years after Pasteur's initial work. The idea of applying the pasteurization method to milk was first commercially tested in Germany in
1880, but was focused solely upon the preservation of milk in order to
improve its shelf life, rather than its health characteristics [17]. The first
recommendation to employ pasteurization to improve the health-related
properties of milk came from the German chemist Soxhlet in 1886. Even
before it was generally accepted that milk carried diseases such as
diphtheria, typhoid, tuberculosis and scarlet fever, both Soxhlet and the

American pediatrician, Jacobi, advocated only heat-treated milk for infant feeding [18].

A debate then evolved around the value of pasteurization compared to sterilization (extended boiling), with Europeans generally favoring the latter technique for infant feeding. Additional work was carried out on the canned sterilization and condensation of milk, but it became obvious that these products had flavors and consistencies so different from fresh milk that they would never be accepted for general consumption. The eventual success of pasteurized milk occurred in America when it was demonstrated that full boiling was neither necessary nor desirable. Pasteurized milk was safe, practical and fit the needs of most urban consumers. It was very similar in taste and color to fresh milk and required no change in consumption or cooking habits. Even though these advantages were obvious, it still was a struggle to establish pasteurized milk on the market. The defensive posturing of the dairy industry, together with the outspoken remonstrations of conservatives who espoused the peerless qualities of natural, raw milk, significantly delayed the adoption and commercialization of the new methods.

"I DRINK TO THE GENERAL DEATH OF THE WHOLE TABLE"

FIGURE 2.4 American Medical Association criticism of milk. The caption reads "I drink to the general death of the whole table" [18].

FORMS OF CERTIFICATES

FIGURE 2.5 Certified milk seals [18].

As milk became more and more associated with the transmission of disease, however, public health officials and the medical community became more direct in their condemnation of the dairy industry and its products. These criticisms varied from harsh, graphic associations of raw milk and death to more sober endorsements of breast-feeding (Figures 2.4 and 2.6) [18].

The dairy industry soon realized that the historic belief in the purity of raw milk was becoming passé and, in response to growing public criticism, slowly began to clean up its act. In collaboration with state medical associations, certain dairies made an effort to place their operations under a much higher degree of hygienic control. Much greater attention was paid to cleanliness of animals, barns and milking utensils. Workers had to be healthy, and the operating procedures were subject to rigid inspection. Milk which was produced under the specified conditions received the endorsement of the state medical milk commissions and was categorized as *certified* milk (Figure 2.5). Certified milk was considered to be the purest and safest raw milk possible to produce.

Even though certified milk was a great step forward in the cleanliness and hygienic quality of milk, it quickly became evident that the procedures involved were no guarantee against milkborne infections. There was no step anywhere in the process to actually destroy any contaminating bacteria. Certified milk was found to be responsible for several outbreaks of diphtheria, tuberculosis and scarlet fever. In some cases, the infections were traced to dairy employees who carried the diseases, but had no outward signs of suffering from them. As a result, the notion that raw milk was dangerous, whether certified or not, began to gain ground. The growing public awareness of the risks of milkborne disease, pointed to the need for new technology. The weight of evidence began to tip the balance away from the old belief that fresh, raw milk was the best milk.

This change in attitude toward raw milk brought forth other approaches to the problem of milk safety, aside from pasteurization.

FIGURE 2.6 Chicago Health Department's recommendation to mothers [18].

Among the most curious methods proposed were the chemical treatments of milk. Some of the chemicals recommended were hydrogen peroxide, salicylic and benzoic acids, and potassium dichromate. The substances which were often employed commercially were borax, boracic acid and formaldehyde. Such compounds were sold to dairies under such trade names as *Freezine* or *Aseptine* [18]. All this was in an effort to get around the need to pasteurize.

Pasteurization Finally Succeeds

The torch for pasteurization was firmly taken up in the United States by a New York philanthropist, Nathan Straus, who was shocked at the mortality statistics of children fed raw milk. He proceeded to establish milk depots all over New York City where the heat-treated (pasteurized) milk was made available [19]. He also recommended the home pasteurization of milk, if it was not available from the local dairy. The Straus Home Pasteurizer is very similar to the apparatus used by so many mothers today to ensure that their baby's formula is safe to drink.

The early history of pasteurized milk was fraught with difficulty. The process was criticized for destroying the "essential quality" of milk. But, because the practical problems of providing an adequate and reliable supply of safe milk to consumers in large cities was an almost insurmountable technical barrier, milk pasteurization was often done secretly. In New York City, the secret pasteurization of milk was outlawed in 1906, and a city ordinance required that all pasteurized milk be so labeled. Demands were made by many to devise tests to determine if milk was pasteurized. People who believed in the local farmer, or even the idea of the family cow, felt that raw, untreated milk was better and posed no health risk. Despite the growing public health information regarding the potential for milk to transmit diseases such as tuberculosis, many influential consumers still believed raw milk to be the best. Many people still believe it.

In their book *Milk Pasteurization* [19], Hall and Trout list the complaints and objections that were originally made against pasteurization. These complaints were placed into five separate categories. Some of these complaints are reproduced in Table 2.1. As will be seen later, many of these complaints have a familiar ring to them.

Thus, the often emotional debate over the risks and benefits of pasteurization long delayed its introduction. The overwhelming body of

Table 2.1 Original objections to pasteurization [19].

A) Sanitation 1. Pasteurization may be used to mask low-quality milk. 2. Heat destroys great numbers of bacteria in milk and thus conceals the evidence of dirt. 3. Pasteurization promotes carelessness and discourages the efforts to produce clean milk. 4. Pasteurization would remove the incentive for producers to deliver clean milk. 5. Pasteurization is an excuse for the sale of dirty milk. **B) Physical and Bacteriological Quality** 1. Pasteurization influences the composition of milk. 2. Pasteurization destroys the healthy lactic acid bacteria in milk, and pasteurized milk goes putrid instead of sour. 3. Pasteurization favors the growth of bacteria in milk. 4. Pasteurization destroys beneficent enzymes, antibodies, and hormones, and takes the "life" out of milk. **C) Economics** 1. Pasteurization legalizes the right to sell stale milk. 2. Pasteurization is not necessary in a country where milk goes directly and promptly from producer to consumer. 3. Pasteurization will increase the price of milk.	4. There are always some people who "demand raw milk." 5. If pasteurization is required many small raw milk dealers will either have to go to the expense of buying pasteurizing apparatus or go out of business. **D) Nutrition** 1. Pasteurization impairs the flavor of milk. 2. Pasteurization significantly lowers the nutritive value of milk. 3. Children and invalids thrive better on raw milk. 4. Infants do not develop well on pasteurized milk. 5. Raw milk is better than no milk. **E) Public Health and Safety** 1. Pasteurization fails to destroy bacterial toxins in milk. 2. Imperfectly pasteurized milk is worse than raw milk. 3. Pasteurization, by eliminating tuberculosis of bovine origin in early life, would lead to an increase in pulmonary tuberculosis in adult life. 4. Pasteurization is unnecessary, because raw milk does not give rise to tuberculosis. 5. Pasteurization gives rise to a false sense of security. 6. It is wrong to interfere in any way with Nature's perfect food. 7. Pasteurization would lead to an increase in infant mortality.

scientific knowledge clearly demonstrated the benefits of pasteurization in the extension of shelf life and the prevention of foodborne diseases [20]. There is little doubt that the spread of this knowledge through the popular press provided both the pressure and incentive for manufacturers and legislators to finally ensure that pasteurized milk was made available to the public. For the first time in history, the food, or more correctly, the dairy industry was able to provide consumers with products that were virtually as tasty and nutritious as the fresh, raw product, but without the risks associated with their spoilage. By the 1920s, pasteurized milk was common throughout the United States and Canada and was compulsory in most large cities. Unfortunately, the benefits it held for the reduction of disease took considerably longer to be realized in most other parts of the world.

Some Countries Still Do Not Have Pasteurized Milk

In 1939, Dr. G. S. Wilson, professor of bacteriology at the University of London, was given the task by the British government to review and report on all the evidence available on the issue of pasteurization. In the preface to his publication he wrote:

> . . . it is useless for men of science to expect that a careful presentation of the facts will make any impression on the minds of those who, whether from unreasonable conviction or from a repressed suspicion of the weakness of their own case, are prepared to resort to half-truths, unverifiable assertions, or, as experience has shown, actual falsehoods in support of their contentions. No matter how clear and well documented the evidence may be, there are always those whom Shakespeare must have had in mind when he made Cicero say:
>
> > "Men may construe things after their fashion,
> > Clean from the purpose of the things themselves."
>
> Much of the current prejudice is doubtless due to lack of education and to false propaganda. The majority of persons who distrust or oppose pasteurization do so because their information is incomplete. Many of them have a partial knowledge of the process, which they may dislike, but few of them are thoroughly acquainted with all the relevant facts, on which alone a rational opinion of its merits and demerits can be based. [21]

Wilson went on to highlight case after case, demonstrating the total elimination of milkborne disease in all areas where pasteurization was made compulsory. He then went on to reveal that many of the complaints against pasteurized milk were patently false. He described the situation in 1933 when the city of Manchester was promoting a bill containing compulsory pasteurization legislation and its well-organized opponents placarded the city with the slogan "Pasteurized milk will kill your babies."

By 1942, when Wilson's book was actually published, the situation regarding milkborne disease in Britain had not changed at all. People, young and old, were still dying unnecessarily from milkborne diseases. When pasteurization finally became a firm reality in the marketplace, no one took responsibility for all the unnecessary sickness, misery and death caused by the deliberate delay of its utilization. Today, in order to allow freedom of consumer choice, the sale of raw milk is still permitted in England and Wales, and it still continues to be implicated in outbreaks of foodborne diseases. Scientists consider this milk to be an unacceptable threat to human health, particularly since the microbiological quality is verified by tests that do not detect pathogenic bacteria [22]. It is therefore an invisible time bomb.

One of the last countries in Europe to enact mandatory pasteurization laws was Scotland. Prior to the enactment of this legislation in 1983, the rate of milkborne salmonellosis in Scotland was the highest in Europe. A year after pasteurization was made compulsory, Scotland's rate was one of the lowest. A study was then carried out on the remaining incidence of milkborne salmonellosis in the three-year period following compulsory pasteurization. Of the fifteen outbreaks that occurred, all were in the rural farming communities and none in the general urban population. This was attributed to the fact that milk consumed in the remote farming districts was exempt from the compulsory pasteurization legislation that applied to the rest of the country [23].

Despite all the evidence, there are still those who wish to consume raw, unpasteurized milk. If they are willing to take the risk, to play Salmonella roulette for personal, political, sociological or nostalgic reasons, they should be allowed to do so but, fortunately, they can no longer compel others to do so. Unfortunately, food irradiation has not reached the same state of rational judgment and consumer accessibility.

The Relationship between Pasteurization and Food Irradiation

The historic coincidence in timing between the work on pasteurization and irradiation was not accidental. Once the relationship between spoilage and disease was known, efforts were put in motion to eliminate both menaces. The use of heat was natural for liquids, because it could be evenly distributed simply by mixing or stirring. The process was even made continuous simply by pumping raw fluid milk in one end of a heated tube and collecting the pasteurized product at the other end.

Unfortunately, the problem was not quite so straightforward in the case of solid foods. Heat cannot be transferred and distributed so easily through solids. You cannot blend up a whole chicken. If you want to use heat to kill *Salmonellae* in poultry, you have to cook them completely in order to ensure that any bacteria in the geometric center of the birds are also destroyed. If you want to destroy the sprouting enzymes in potatoes with heat, you have to cook them thoroughly for the same physical reason. This alters the product to such an extent as to make it unmarketable to the general public. Therefore conventional heating techniques could not be applied to solid foods in the same way as they were to liquids. However, any technology which ultimately allowed energy to be transferred quickly and evenly throughout a solid food without appreciably altering it, could finally provide the same degree of freedom from foodborne diseases which pasteurization does for liquid foods.

Although the food irradiation technique worked well on solid foods almost 100 years ago, it was simply not feasible at the time because there were no cheap and practical irradiation sources available. Had there been, there is no doubt irradiation would have become as well-established as pasteurization. As it happened, when practical irradiation sources finally became available and economically feasible, the public perception of the process was more closely identified with military, rather than peaceful, uses of nuclear energy.

This false perception was compounded by the misinformation spread about the effects of irradiation on food. Any issue involving the word *radiation* was easily placed out of context, and provided anti-irradiation advocates years of controversy and publicity. The ease of relating this new technology to the inherent fear of radiation poisoning afforded an ideal opportunity for these groups to foster and to feed a phobia. The built-in anxiety that people had of ''things nuclear'' was easily exploited. It meant that hard, factual information would not be necessary to challenge the scientifically established evidence that irradiated foods were safe. Hard evidence isn't necessary to reinforce an innate fear.

Evidence was presented which was biased, slipshod and otherwise totally unable to withstand scientific scrutiny. The issue was no longer safer or better food—the issue was *nuclear*, it was *environment*—food had to be "green" because "green" was in. It was no longer health, it was politics. And it was no longer science, it was nostalgia. Consequently, the voices of these "experts" and "representatives" of the common people were heard above those of the recognized scientific authorities. They were given more credibility than acknowledged experts because we all have a natural fear of the unknown. The fact that scientists gave an assurance of safety meant little to the understanding of an issue that was largely misunderstood. Wilson's 1939 statement applies just as aptly to food irradiation today as it did then to pasteurization.

It is not difficult to misrepresent facts and information. Considering that free radicals and thermolytic products are formed during pasteurization, there is little doubt that if attempts were made to introduce pasteurization today, it would probably suffer much the same fate as food irradiation—with the very same arguments (see Table 2.1).

Irradiated Milk

The milk industry provides us with another interesting parallel to food irradiation. The process of milk irradiation was used during the 1930s in order to increase the vitamin D content of milk. By 1935, there were an estimated 35,000,000 consumers of irradiated milk in North America. In addition, over half the entire production of evaporated milk, which represented a million tons of fluid milk, was irradiated [5]. Even Red Cross food packages for prisoners of war included irradiated milk (see Figure 2.7) [186]. The difference between milk irradiation of the 1930s and food irradiation of today is only the wavelength of the electromagnetic spectrum chosen to carry out the work. Milk irradiation was carried out using ultraviolet (UV) light as the source of ionizing energy rather than γ-rays.

The idea to use this type of irradiation on milk to increase its vitamin D activity derived from the natural action of sunlight upon plants. Sunshine does the same to our skin. The UV radiation in sunlight works by a mechanism of ionization and free radical formation to change a natural steroid called 7-dehydrocholesterol into vitamin D. Rickets is a disease of infancy and childhood, where bone hardening is delayed because calcium metabolism is deficient due to a lack of vitamin D. In view of the common occurrence of rickets, it was felt there was a

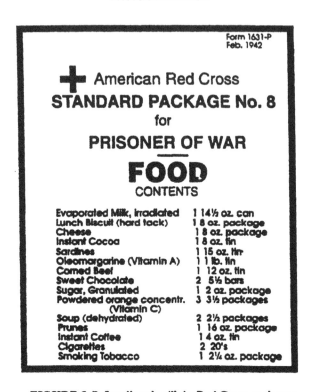

FIGURE 2.7 Irradiated milk in Red Cross packages.

legitimate need for a convenient source of an antirachitic (anti-rickets) factor in foods commonly consumed by children. Although it was theoretically possible to add vitamin D to milk, the available sources were neither cheap nor convenient. Earlier work carried out in the 1920s revealed that certain foods or their original sources, after exposure to light energy of particular wavelengths, had a greater ability to prevent rickets [24,25].

Research at the time indicated that vitamins B and C might be partially destroyed as a result of ultraviolet irradiation, but it was judged that the overall benefits of increased vitamin D outweighed any possible negatives resulting from minor losses of the other vitamins. There were also some flavor problems initially, but these were overcome to a point where they were not considered a problem for consumers. (The same flavor problems, which are due to oxidation and not souring, occasionally occur today when bottled milk is left too long on the doorstep and exposed to the sun, even if it is cold outside.) Not surprisingly, irradiation of milk with γ-rays comes up against the same flavor problems as with ultraviolet rays and have been solved in the same way. In 1930, inert nitrogen gas was used to eliminate oxidation resulting from ultraviolet irradiation [5], and the same approach was used almost sixty years later for the irradiation of milk with γ-rays [26].

As vitamin D eventually became cheaper, its addition to milk replaced the use of UV irradiation. Ultraviolet irradiation of milk is still carried out in certain dairy plants, but this is more for bacterial control than for increasing the levels of vitamin D.

What then, are some of the lessons to be learned from the parallels between ultraviolet irradiation and irradiation with γ-rays?

- It is clear that the term *irradiation* refers to a process of exposure to all wavelengths in the electromagnetic spectrum and not just to nuclear radiation.
- Many wavelengths in the electromagnetic spectrum can cause ionization, free radical formation and molecular changes.
- Consumers do not fear the term *irradiation* per se, unless it is erroneously linked with the atomic bomb or Chernobyl.
- Millions of children were given a greater protection from rickets even though there were some other vitamin losses associated with an irradiation process.

The case of irradiated milk is interesting, because it allows a strikingly curious example of a technology which is as close to food irradiation as one can get. The decisions to employ this technology were based upon rational scientific studies and nothing else. It served its purpose until a better technology came along to replace it. Perhaps many readers of this book would have been spared bowed legs if irradiated milk were available to them when they were young. Unfortunately, they were not given the option to have greater protection from disease. And perhaps that is the greatest parallel with food irradiation.

CHAPTER 3

Foodborne Diseases

Perception of the Problem

Recent data by health professionals place the estimated annual number of cases of intestinal infectious diseases in the United States at 99 million [27]. This means that more than one in three Americans suffer from some form of gastroenteritis every year. A major factor in these figures is foodborne disease. Since the food system in the U.S. is technically advanced, and regulations are strictly enforced, it is very possible that the rate of such incidents in other countries is even higher. If the global figures for foodborne diseases were to be estimated on a yearly basis, they would be staggering.

Although it may not provide any comfort, it should be clear that the more recent appreciation of the seriousness of various food problems does not necessarily mean they are new. In most cases, it simply means they were not well-identified previously. In the case of foodborne diseases, new or improved methods for the detection of microorganisms and their toxins have uncovered problems which were formerly hidden from us. Contrary to what some may think, a problem that is unknown or ignored is no less of a problem. On the contrary, the documented knowledge of a problem is the first step in its resolution.

Consumer knowledge of foodborne disease is extremely limited. Despite their significant implications on the general health of people the world over, foodborne diseases are not an issue of great public concern. This is unfortunate. For the past two decades, the U.S. Food and Drug Administration has ranked food-related risks in the following order of importance [28].

(1) Foodborne disease

(2) Malnutrition

(3) Environmental contaminants

(4) Naturally occurring toxicants

(5) Pesticide residues

(6) Food additives

This ranking is generally agreed upon by most public health and related agencies throughout the world. In contrast, consumers (particularly those from developed countries) place food additives and pesticide residues at the top of the list, and foodborne diseases near the bottom.

In a fairly recent British Government survey on consumer concerns of health risks, smoking, pollution and additives ranked as the first three of six issues [29]. Foodborne diseases did not even rank! Why is there such a difference in the perception of relative risks between professionals working in the food and health sector on the one hand, and consumers on the other?

Is it that consumers are simply uninformed? Or are they misinformed? Is the issue of foodborne disease routinely swept under the rug because it is a repulsive subject about which we feel totally helpless? Most probably, it's a combination of all these factors that allows us to overlook the issue. This is a terrible pity, since we are all so vulnerable to what is essentially a rather simple and controllable biological phenomenon.

The fact is that most incidents of foodborne disease are largely avoidable, on the part of the consumer, as well as the perpetrator. The vast majority of outbreaks occur in catering establishments, in street kitchens, and in the home. Yet, a little understanding and appreciation of the subject would relieve immeasurable grief, discomfort and misery from so many of our lives. If consumers had an adequate knowledge of the issues, they would not take so many stomachaches, 24-hour flus or bouts of nausea for granted.

Some of the more infamous outbreaks have involved thousands of people and many deaths. The largest *Salmonella* outbreak was estimated to have affected 165,000 people in an area covering six midwestern states in the U.S. [30]. Other large outbreaks have occurred in Sweden, Canada, Netherlands, France and other countries. In one incident in Scotland, a single can of corned beef was implicated in the infection of over 500 people. This was not due to the thrifty nature of the Scots, but rather to the machine that was used to slice the contaminated corned beef, and continued to be used to cut many other meats, and thus spread the infection dramatically. Even chocolate was implicated in the *Salmonella* poisoning of over 200 people in Canada [31]. In fact, the only recorded grounding of the Concorde transatlantic service was due to food poisoning.

All strata of society are affected, rich and poor, manual laborers and professionals alike. There was a rather interesting report published recently on an outbreak of *Salmonella* poisoning among the delegates of a Welsh medical conference. At least 196 out of the 266 delegates who attended the buffet were seriously affected [32]. There is no telling how many ill-fated kitchen staff, waiters and busboys also stole a nibble of

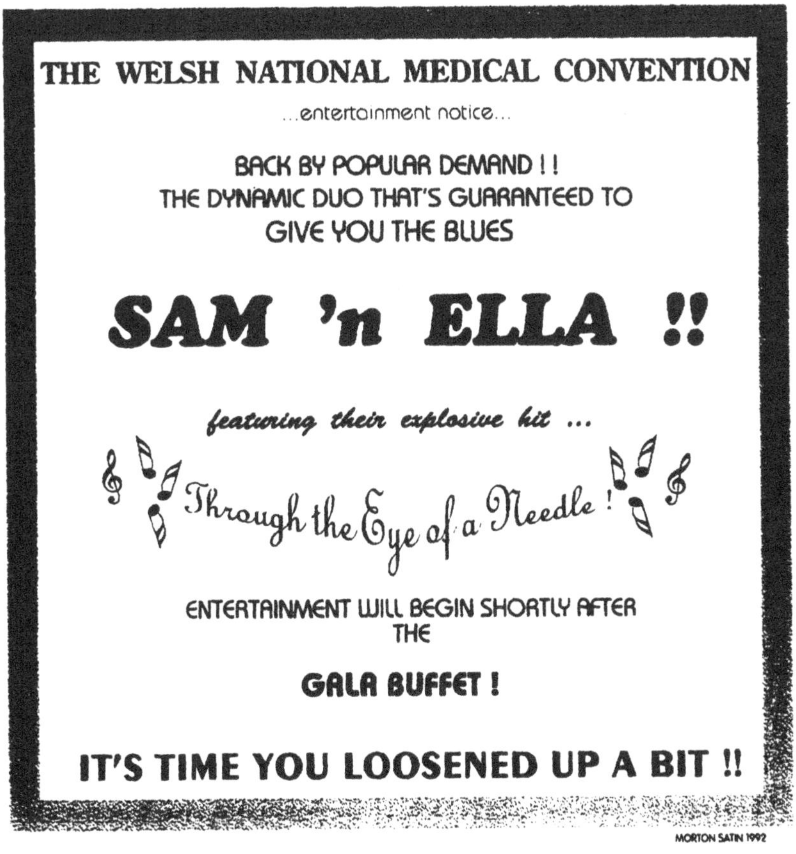

FIGURE 3.1 Medical convention notice?

the suspected chicken, or cooked beef, or chicken liver paté, or wholemeal bread crumbs that were all later found to be *Salmonella*-positive by the weakened, but vengeful doctors ("It was the paté!" they cried). In fact, the problem of *Salmonella* in poultry has become so common, that many Scottish hospitals no longer serve chicken [33]. No doubt, this move will attract many medical conferences away from Wales in the future.

The startling figures on *Salmonella* outbreaks don't necessarily mean that the food systems in the countries mentioned are worse than in other countries. It means that the systems required for the full diagnosis, disclosure and reporting of these problems are well developed. If all countries had similar analytical and reporting systems, the foodborne problems revealed would demonstrate the same sort of magnitude. In fact, it is the countries that do not publicly report these incidents which should be suspect, as there is no evidence to demonstrate that their own particular food systems are any different.

Foodborne diseases are not simply minor inconveniences. They are real diseases and should never be lightly considered. While in most instances the symptoms of these diseases are not long lasting, in many cases they can end in permanent debilitating effects, and even death, as a result of complications. Although they are often the subject of humor, foodborne diseases and their symptoms are definitely not a joke.

In presenting information on the seriousness of foodborne diseases, it is important to ensure that the issue is neither exaggerated nor taken out of context. The purpose is not to shock, but rather to inform. Still, the ubiquity and significance of foodborne diseases must be highlighted, and their causes described.

Terminology

In order to get a better picture of what they are, it is best to start with a few definitions.

A *foodborne disease* is simply any illness that results from eating foods that contain toxins or pathogenic microorganisms.

Toxin is another word for poison, and a *pathogen* is any microorganism capable of producing a disease or sickness.

Foodborne diseases fall into three general categories [34]:

- *Food infections* are caused by the consumption of foods which contain sufficient numbers of pathogenic microorganisms to colonize the victim's intestinal tract, and thus cause the symptoms and damage common to that organism. These are the most common forms of foodborne diseases. Typical examples are salmonellosis, shigellosis and campylobacter enteritis.

- *Food intoxication* results from the ingestion of poisons. The most common toxins result from the growth of microorganisms in foods. Chemical toxins (the majority of which are from natural sources), account for a very minor percentage of foodborne disease cases [35]. The most common examples of food intoxication are staphylococcal food poisoning and botulism.
- *Food toxicoinfections* are, as their name implies, a blend of food infections and intoxications. The ingested food contains sufficient numbers of pathogens to infect the victim's intestinal tract and to produce toxins which result in symptoms of the disease. Examples of typical microorganisms are *Clostridium perfringens*, *Escherichia coli* and *Vibrio cholerae*.

The term *microorganism* should also be clarified. Microorganisms are organisms that are so small (0.1 μ – 100 μ[1]), they are generally invisible to the naked eye. They are made up of bacteria, yeast, mold, protozoa and viruses. Although only bacteria, yeast and molds produce toxins, all classes of microorganisms are capable of causing infections.

The last concept to understand is that of an *offending* organism. Offending organisms or microorganisms are those that cause harm or damage. All the technologies employed to control spoilage or pathogenic microorganisms do not actually remove them. This is physically impossible (unless you are filtering liquids, such as water, through a micropore filter that blocks all organisms except some viruses). All related food technologies simply seek to remove the organism's offensiveness rather than the actual organism itself. Pasteurization does not remove any bacteria, nor does it kill them all. It merely kills enough of them to give milk a practical shelf life and reduces the amount of pathogens to a level that we can all safely tolerate. However, if infective microorganisms are no longer able to infect, they are no longer offensive. Because infection usually requires exposure to a particular minimum number of bacteria, depending upon the pathogen in question, any process that reduces their number below that minimum is sufficient to provide an acceptable

[1]1 μ (micron) = 1/10,000 of a centimeter or 1/25,000 of an inch.

degree of safety. Of course, there is also the alternative of sterilized foods, in which, by definition, all microorganisms are killed (but not removed). However, sterilized foods usually suffer from major changes in eating characteristics (canning, for example). The same situation exists in the case of parasites. If parasites are damaged to the extent that they can no longer parasitize, they too are no longer offending organisms. It is extremely difficult to eliminate naturally occurring parasites from the environment. Until that is accomplished, technologies must be employed to ensure that they cannot reproduce and cause harm. The concept of an offending or nonoffending microorganism should be clearly understood if consumers are to get an accurate perspective of the issue of foodborne disease. Once destroyed, or rendered inoffensive, one bacteria is much the same as another, and pathogens thus treated are no worse than the loads of *Lactobacilli* found in pasteurized yogurt.

Our Awareness of Foodborne Disease

For most people, the unpleasant experience of foodborne disease lasts only a few days. However, a significant number of cases can be very serious, and cause long-term chronic gastric problems and other related difficulties. The very young, as well as older people, are particularly at risk from foodborne diseases. There have been some rather spectacular, well-publicized outbreaks of food-related poisonings or infections but, unfortunately, most incidents of foodborne illness go almost totally unreported. There are a number of reasons for this. In many cases, people simply don't think the symptoms are serious enough to warrant the time and expense of consulting a physician and furnishing samples for analysis. The symptoms, typically diarrhea and stomachache, are not uncommon, and people usually feel they will improve without resorting to a doctor. Even if they do see a doctor, unless blood or stool samples are taken and analyzed, there is no way of knowing the exact cause. Consequently, the incident is never reported. It does not take much imagination to recall the number of times this has happened to us all. Of course, not all cases of diarrhea are due to foodborne diseases. However, a very significant number of such incidents definitely are.

It has been variously estimated by health professionals that anywhere from 1 in 25 to as few as 1 in 5,800 of the true number of cases are actually reported [36]. When a reasonable estimate is made of the health

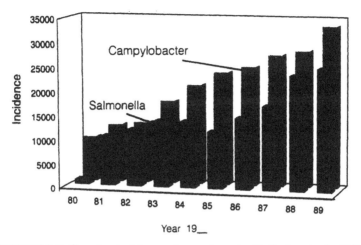

FIGURE 3.2 Growing incidence of gastrointestinal infections in the U.K.

and economic effects of foodborne diseases, it becomes readily apparent why food and health experts consider them to be the greatest of all food-related risks.

In order to get some perspective on the problem, it is worth looking at the increase in reported incidence of gastrointestinal infections in the U.K. over the period 1980–1989 (see Figure 3.2) [37]. Although all these may not be due solely to foodborne infections, the dramatic upward trend is clear. It is evident that, as our food system gets more complex

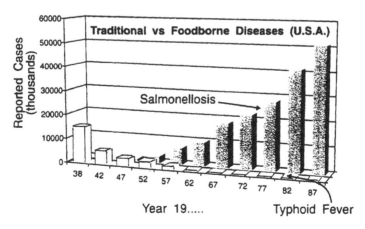

FIGURE 3.3 Fifty-year U.S. trend in salmonellosis and typhoid fever.

in order to feed an ever-growing population in urban centers, the requirement for more effective and appropriate technologies becomes more critical.

Another aspect worth examining is the relationship of foodborne to other major diseases, such as typhoid fever. The data over a fifty-year period are given in Figure 3.3 [38]. Thus, although modern medicine and hygiene have been successful in arresting certain diseases, they have not been able to stem the dramatic and consistent rise in foodborne diseases.

Prior to looking into those aspects of the issue, a very basic knowledge of the key diseases and the agents responsible is necessary to better understand the topic. The diseases and their causative agents will be described in the generally agreed list of their relative importance. Of course, the list may vary significantly from country to country, or even with particular lifestyles.

Bacterial Diseases

Salmonellosis

Salmonellosis is one of the most frequent causes of recorded foodborne disease. It results from an infection of a bacteria called *Salmonella*. This bacteria does not owe its name to the fish, but rather to the American bacteriologist, D. E. Salmon, who characterized hog cholera over 100 years ago. In fact, the mistaken association of this disease with the fish was so great that a senator from the salmon-fishing state of Washington proposed a bill for Congress to change the name of the bacteria to *Sanella* and the disease to sanellosis.

Depending upon the type of *Salmonella* bacteria involved, the symptoms and severity of disease can vary greatly [39]. The most hazardous type is *S. typhi*,[2] the bacteria responsible for typhoid fever. In developed countries, this particularly dangerous variety is not common, and the incidence of typhoid fever is generally low. But it does occur. It is a different story in developing countries where the sanitation systems are not sufficient to handle it. The same can be said for any locale where the sanitation system breaks down, particularly after war, earthquakes, hurricanes, etc.

[2]The standard way of describing bacteria, once the genus or species name (in this case *Salmonella*) is given, is to abbreviate it with a capital letter and spell out the type, usually in italics. Thus *Salmonella typhi* becomes *S. typhi*.

The spread of *S. typhi* is fairly straightforward, and similar in many ways to other foodborne diseases. Infected people shed the bacteria in their feces. The *Salmonella* enters the food and water system through direct unsanitary contact, or through untreated sewage. Reinfection is thus virtually guaranteed.

Typhoid fever is a very serious disease. It is an example of a relatively rare case where a foodborne pathogen can penetrate the intestine, enter the bloodstream and start to multiply. The symptoms include high fever, headache, vomiting and diarrhea. These bacteria can actually enter the very cells in the body (macrophages) that are supposed to defend us against them. Once the *S. typhi* enter these cells, they are very resistant to antibiotics, and repeated therapy is required. Almost any food can become the vehicle for transmission of this disease. The most famous case on record is Typhoid Mary, who was a chronic typhoid carrier, although she herself did not experience severe symptoms [40]. She was a carrier, she had poor personal hygiene habits (a common occurrence), and she worked as a cook — a lethal combination of factors which resulted in multiple outbreaks of typhoid fever. Fortunately, typhoid fever can be virtually eliminated through scientifically applied control measures to our sanitation systems.

Under the right conditions, several other types of *Salmonella* bacteria can produce similar symptoms, often referred to as paratyphoid fever. It is also generically referred to as enteric fever, due to an acute inflammation of the intestine. Two varieties of bacteria involved are *S. paratyphi* and *S. hirschfeldii*.

By far, the most prevalent form of *Salmonella* infection is caused by the large number of remaining types. They are found in all foods such as meat, seafood, dairy products and even fruit and vegetables [41]. *Salmonella* bacteria are so common on raw meat and poultry that it has been suggested they be considered part of the normal microflora of these products [42]. The typical symptoms include severe diarrhea, dehydration, fever, vomiting, headaches and abdominal cramps. Depending upon the individual, the symptoms usually last from 1−5 days. Typical culprits are *S. typhimurium, S. enteritides, S. heidelberg, S. newport, S. montevideo, S. dublin, S. meunchen, S. manhattan, S. havana, S. orianenburg,* and *S. saint paul.* The various types read like an airline travel guide. This should not come as a surprise because, by now, it should be obvious that *Salmonella* can really get around.

The majority of *Salmonella* infections are derived from the consumption of food and water. Beef, chicken, turkey, pork, fish, seafood, dairy products and eggs have all been identified as carriers of this pathogen. The bacteria infects our food plants and animals by a variety of routes, including unprocessed sewage or manure used as fertilizer, in animal feeds and in the dust and dirt associated with animal wastes. The spread of the bacteria has been so great, that it is considered a *geonosis*, which means it can come from soil and water, as well as animals. In the case of fish and seafood, it was previously thought that the presence of *Salmonella* was a postcatch phenomenon due to unsanitary handling, but it is now accepted that salt-tolerant varieties of the bacteria are becoming more predominant. This has been attributed to the increased dumping of unprocessed urban wastes into the marine environment. It can be expected that this continued activity will result in many more marine-resistant strains of *Salmonella* as time goes by. It may also be predictive of a potential tolerance of *Salmonella* to the proposed trisodium phosphate treatment.

Prepared products, such as cream-filled baked goods, and Mexican, Chinese and other popular ethnic foods, have also been involved in *Salmonella* outbreaks. Even fruits and vegetables can harbor *Salmonella*, particularly when the new vogue of using natural, unsterilized manure as a fertilizer, is employed. The food most commonly recognized as a major source of this pathogen is poultry and its products. Internationally, the incidence of *Salmonella*-positive poultry varies up and down between 15–90%. This applies to all countries. The poultry industry is naturally very sensitive to this issue and does whatever it can to keep levels at a minimum, but with the technology currently available, the ability to eliminate *Salmonella* is limited [43]. There are many self-styled experts who think differently – but they have thus far offered no viable solutions to the problem. If they did, the poultry industry would be quite eager to use them. The situation is not unlike that of certified milk, where, despite all precautions taken, there was no step anywhere in the process to ensure the actual destruction of bacteria.

The risk of salmonellosis could be substantially reduced if consumers were aware of the high incidence in beef and poultry, and took appropriate measures to prevent infection. *Salmonella* is very heat sensitive, and is easily killed during conventional cooking. People seldom get the infection from eating properly cooked poultry products. However, recent studies were carried out on the survival of *Salmonella* in chicken cooked in a conventional electric oven, a convection microwave oven and a regular microwave oven. The results clearly indicated that cooking poultry in a regular microwave oven to the highest internal temperatures recommended, did not ensure complete destruction of *Salmonella* [44,45]. Although the correct internal temperature may be reached, the full time/temperature profile required to kill certain pathogens is not achieved. It is therefore critically important to heed the manufacturer's instructions for leaving the poultry to rest, hot, for a period after microwaving, in order to continue the cooking action to destroy pathogens. It would also be useful if the manufacturers would explain to consumers the reason for leaving the poultry to rest.

Most *Salmonella* infections result from cross-contamination in the kitchen or in the supermarket. Try to imagine preparing a chicken at home. The first thing one does is to remove the packaging. Hands are exposed, the kitchen counter is exposed. Perhaps a knife was used to open the package. The poultry is then prepared. Again, hands, counter, utensils, apron, plates, pots, sink, etc., are exposed. Poultry, like all other animal products, contains fat. Fat can protect *Salmonella*. In order to get rid of contamination, careful washing with soap is an absolute minimum—at every stage! If not, whatever one touches can become cross-contaminated.

The same situation exists at the supermarket. Packaged poultry often has juices leaking out the bottom. (All such packages should be reported to the store manager.) If the leaking juices contain *Salmonella*, then everything else touched by the customer—fruit, vegetables, cheese, etc.—will be cross-contaminated with the same bacteria. The problem is that these cross-contaminated products are not cooked afterwards to kill the bacteria.

Even with all the precautions and hygienic measures in place, it is still very difficult to remove the possibilities of infection or cross-contamination once *Salmonella* is present, because the bacteria is so persistent. This bacteria has the ability to stick to its host, despite heavy washing with soap.

Poultry is not the only source of this microbe. All other meat and fish products afford the same sort of *Salmonella* risk. When all sources of *Salmonella* food poisoning incidents are combined, and an economic value is attributed to this disease, the overall impact is very significant indeed.

Clear evidence that the problem has been swept under the rug, for whatever reason, is that the incidence of *Salmonella* outbreaks is increasing, not decreasing. Look once again at Figure 3.3, which compares the incidence of salmonellosis to typhoid fever. What is even worse is that consumers have absolutely no way of knowing which products may be contaminated with *Salmonella*. Although it is known that food irradiation can be very effective in reducing the risk of foodborne *Salmonella* infection, consumers have not been given the option to choose *Salmonella*-reduced food products, even though they might be easily available. It is a case where the knowledge and tools proven capable of doing the job are simply not being put to use.

Perhaps the biggest surprise is that pathogen-contaminated food is actually permitted, when there is a proven, safe method available to prevent it. One would think that the century-long experience with pasteurization would serve as an example of the deliberate procrastination that can prevent the introduction of improvements to our food system. But it has not. *Salmonella* poisoning is not an unavoidable risk. Unfortunately, legislators, manufacturers and retailers are not giving consumers a chance to avoid the risks of Salmonella roulette.

Staphylococcal poisoning

As previously described, the terms *food poisoning* and *food infection* are not really interchangeable. In the latter case, a microorganism is ingested with the food, it multiplies in the gastrointestinal tract, then attacks the tissues, which results in the disease. In food poisoning, a toxin is present in the food before it is consumed, and results in the intoxication when it is eaten. Destroying the offensive microorganism does not make the food less toxic. Staphylococcal poisoning is the most frequent cause of food poisoning in many countries.

As the name suggests, the causative factor in this disease is a toxin produced by the *Staphylococcus* bacteria. The most common type involved is *S. aureus*, better known as the common hospital *staph*. The bacteria itself is quite sensitive to heat and other conventional means of control. Fresh, unprepared foods are seldom a source of toxin. (An exception is unpasteurized milk coming from cows with mastitis, an udder infection.) The origin of this foodborne disease is most often the preparation of foods by employees who are infected with large numbers of the bacteria. It is estimated that approximately 50% of all people are carriers of *Staphylococcus* at one time or another. When people handle food, infection comes from hand sores, or by coughing and sneezing (there are usually high numbers of the bacteria in the nasal passages).

Once in the food, the *S. aureus* multiplies and produces a poison called an *enterotoxin* (or intestinal poison). It is the ingestion of this enterotoxin that actually causes the illness. When people eat the contaminated food, they are actually poisoned in the classical sense, and therefore, the symptoms become evident in a very short period of time (from 2–4 hours after eating) [46].

The most common place where contamination occurs is in foodservice and catering establishments. The incoming food may be clean, but contamination can result from mishandling. If these foods are not refrigerated quickly after preparation, then an ideal medium for the growth of *Staphylococcus* and the production of toxin presents itself. Once the toxin is formed, it is stable, and the food becomes a poison looking for a victim. The main symptoms of the disease are vomiting and diarrhea. More recently, however, it is becoming evident that staphylococcal food poisoning does have the potential to exhibit the symptoms of toxic shock syndrome, a condition that could prove fatal [47].

S. aureus occurs in baked goods, minced or chopped pork, beef and poultry products, cheese, eggs, fish, fruits and vegetables. The foods most frequently implicated are the cream or custard fillings of pastries. Potato salad, take-out meals and shellfish have also been involved in transmitting staphylococcal poisoning to consumers.

Foods that are contaminated with staphylococcal enterotoxin cannot be decontaminated. The toxin is destroyed by neither heat nor food irradiation. The production of toxin can therefore only be prevented by ensuring minimal levels of *S. aureus* in the foods at all stages of preparation. Since this bacteria can grow on most foods that are not acidic, cleanliness and refrigeration are the most effective control measures. It is basically a problem of food handler education, training and regular safety analysis for the presence of the toxin [48].

Botulism

Perhaps one of the best-known foodborne diseases is *botulism*, a food intoxication caused by the bacteria *Clostridium botulinum*. The name, *botulism*, owes its derivation to the Latin word for sausage, *botulus*, since the disease was first described in 19th century Germany as *sausage poisoning*. Despite its occurrence in many other foods, the original name has remained with us over the years.

The poison which *C. botulinum* produces is extremely lethal, with as little as a tenth of a gram of food capable of causing botulism. It is so toxic that there have been many stories regarding its use in biological warfare programs. The toxin causes muscle paralysis and is therefore called a neurotoxin. Death can result because the respiratory muscles of the diaphragm no longer function and the victim literally suffocates. Other forms of the disease, such as wound botulism (which is poisoning through a wound infection), and infant botulism are quite rare and will not be discussed here, as they are not considered foodborne.

There is some public misunderstanding about botulism. *Clostridium botulinum* is an *anaerobic* bacteria, which means that it is a bacteria that grows in the absence of air. This is why canning, bottling or preserving of foods under oil, if improperly done, is no guarantee against the disease. Improper preservation also explains why most outbreaks of botulism are actually caused by home preservation, not commercially processed foods. The few commercial outbreaks which occurred have, naturally, received a good deal of publicity. It is unfortunate, however, that the occasions were not taken by the press to remind consumers to be cautious of the even greater danger of improperly prepared home-made products.

Foods that are most often implicated are vegetables and meat products. Particularly troublesome have been home-canned or bottled peppers, beans, spinach and asparagus. Since spores of *C. botulinum* are found everywhere in the environment, particularly in the soil, it is not surprising that plant foods are an excellent potential source. Other foods involved are canned fish, fish eggs, mushrooms, soups and sauces. Home-processed ham, fish, liver paste and venison jerky have also been implicated [49].

Although *C. botulinum* spores are generally quite heat resistant, commercial canning was specifically designed to ensure their destruction. Cans must be processed to a "botch" (botulism) cook at the geometric center of the can. The destruction of *botulism* spores has been the subject of much research and study, and a number of books have been written specifically on the subject. Considering the incredible volume of canned goods that are commercially produced, it is both fortunate and amazing that so few botulism incidents have actually occurred. The rare incidents which have occurred in commercial canning resulted from accidents such as leakage of the can seams. The main reasons for home-processed incidents, on the other hand, are improper temperatures, faulty equipment and poor recipes.

Acidic conditions prevent the germination of spores and the consequent production of toxin. Also, the level of acidity is rather critical. During the period 1899–1975 in the United States, thirty-four out of the thirty-five botulism outbreaks that occurred in supposedly high-acid foods involved homemade products (half of which were tomato-based). When in doubt, it is best that the product be more acidic. Acids commonly used are acetic, lactic and citric acids [50].

While the spores of *C. botulinum* are heat resistant, the botulism toxin itself is rather heat sensitive. Thus, if a food is in any way suspect, boiling it for 10 minutes will inactivate the toxin. A product from a swollen can, or with a poor odor should never be tasted before boiling. To be safe, when in doubt, throw it out.

The new generation of extended-life refrigerated foods may also pose a problem with *C. botulinum*. Such products include soups, sauces, salads, pasta, seafood and meat entrées. Many of these products are hermetically sealed under vacuum or under a modified atmosphere, both of which create ideal anaerobic conditions. Since it has been demonstrated that *Clostridium* can grow and produce toxin under refrigerated conditions, particular care must be taken to ensure the safety of these products. Light cooking prior to refrigeration is not sufficient to kill all spores, and could worsen the problem by killing other nonpathogenic bacteria which might have competed with *C. botulinum* [51]. This also may be the case with food irradiation, although recent work appears to indicate its value in reducing toxin production in these products [52]. The complete control of botulism in this case depends upon the employment of a battery of preventative measures (acidity, heat, salt, etc.) that individually may not be sufficient, but in combination do the job.

Clostridium perfringens

Another *Clostridium* species that causes foodborne disease is *C. perfringens*. This bacteria is also anaerobic, but does not act like *C. botulinum*. Victims usually ingest foods containing large amounts of the bacteria, which then inhabit and flourish in the intestine. When they form spores, they release an enterotoxin that causes the symptoms of stomachache and diarrhea [53]. It is therefore typical of a toxicoinfection. The onset of disease appears to be dependant upon the presence of a heavy load of bacteria. *C. perfringens* does not produce the deadly botulism toxin and the symptoms are usually over in a day or two.

The foods most frequently implicated are beef, chicken, turkey and pork. Spices used in conjunction with these foods are also excellent sources of *C. perfringens*. A recent study on fresh sausages (chorizos) in Argentina revealed that 110 out of 136 samples were positive for *C. perfringens* [54]. Dairy products are another good vehicle for this bacteria.

Campylobacteriosis

During the last decade, the most common type of foodborne bacterial diarrhea and gastroenteritis incidents have resulted from *Campylobacter jejuni*. Despite the fact that little was published about this bacteria before 1980, the recoveries of *Campylobacter* have recently exceeded those of *Salmonella* from gastroenteritis victims [55]. The symptoms include diarrhea, stomachache, vomiting, fever and headache. It may persist for 1–15 days, but is usually over within a week. Because the symptoms are similar to stomach flu, it can easily be mistaken for it. Oddly enough, the height of incidents regularly occurs between May and December, with a peak in the month of July.

Untreated water, raw or improperly pasteurized milk, and under-cooked poultry, seafood and hamburger have all been found to be principal sources of infection. Barbecues are a particular problem, because it is very easy to transmit the bacteria from raw meat to other foods, since so much careless handling is done [56]. *Campylobacter* is present in anywhere from 30–100% of tested poultry, and despite the efforts put in by the industry, it continues to appear routinely in most supermarket chickens [57]. It can be found anywhere, but is chiefly associated with foods of animal origin. Its transmission in milk has even been attributed to wild birds, such as jackdaws, as a result of their aggressive habit of pecking through the tops of milk bottles [58]. As in the case of *Salmonella*, cross-contamination of other foods in the kitchen can occur easily with *Campylobacter* bacteria.

Listeriosis

A pathogen of growing concern is *Listeria monocytogenes*. Although this bacteria was always associated with animal disease in the past, it has more recently been linked to serious diseases in humans. *Listeria* can cause meningitis and encephalitis as well as severe pregnancy or fetal infections. Victims are usually the very young or the very old, and the fatality rate is about 70% if the condition is left untreated [59]. Listeric abortions can occur in the last half of pregnancy and result in stillborn, or acutely ill babies. If born alive, the babies die shortly thereafter [60]. The disease can be transmitted in many ways, including mother to fetus, infant to infant and animal to human [61]. The critical role of food in the transmission of this disease was only discovered recently [62], and it now appears that this is the most common cause of human *Listeriosis*.

The bacteria has been found in a wide range of dairy products, raw vegetables and meat products in most countries. Several studies of meat products sampled in European supermarkets quoted the recovery of *Listeria* to be as high as 80% [63,64], and a recent Australian study on samples picked up at the retail level showed over 42% positive for *Listeria* [65]. Raw, unpasteurized milk, and the soft cheeses that are traditionally made from it, are a growing concern, since it has been clearly demonstrated that *Listeria* continues to thrive well at refrigeration temperatures (4°C). Of additional concern is another Australian study which recently demonstrated the relative ineffectiveness of microwave cooking in destroying *Listeria monocytogenes* in ground beef [66].

The problem should not, however, be taken out of context. According to present evidence, most people exposed to *L. monocytogenes* don't appear to get sick. Normal, healthy individuals who possess no underlying illness are quite resistant to the bacteria, and our defense mechanisms seem to handle it rather well. Because of this, the actual incidence of reported disease outbreaks for *Listeria* are not very high. People whose immune system is not very strong (such as infants, young children, pregnant women, AIDS victims, chemotherapy patients, and the elderly) are at a much greater risk. It is definitely not a bacteria to play roulette with. The consequences of losing the battle with *Listeria monocytogenes* are grave indeed.

Shigellosis

Shigellosis is an important disease in developing countries, as well as in some developed countries where crowded situations occur (such as in schools and institutions). It is also referred to as bacillary dysentery and is a very harsh form of diarrhea. In infants, the fatality rate can be as high as 25%. The microorganism responsible is a small bacillus (rod-shaped bacteria) named *Shigella*, after the Japanese scientist, Shiga, who first associated it with dysentery.

There are several different types of this bacteria, the most common one found in victims of *S. sonnei*. The characteristic symptoms of shigellosis are similar to many other infectious foodborne diseases; e.g., diarrhea, stomachache and vomiting, which can be rather severe. In most cases, the symptoms are usually over in a week or two. Because humans are the chief carriers of the bacteria, infected food handlers are the primary cause of transmission. Almost all foods are susceptible, particularly if they are not cooked just before eating.

Escherichia coli *infections*

Another bacteria which is transmitted in much the same way as *Shigella* is *Escherichia coli*. Although this bacteria is a normal resident of the human intestinal tract, certain types can be pathogenic. The most famous symptoms of this bacteria are familiar to us as *traveler's diarrhea*, *tourista*, or *Montezuma's Revenge*. This often occurs to people who travel from a clean environment, where most foods are prepared in a fairly sanitary manner, to a location where hygiene does not count.

There are four forms of disease associated with the *E. coli* bacteria — *enteropathogenic* (severe diarrhea), *enteroinvasive* (damages the intestinal tissue), *enterotoxigenic* (traveler's diarrhea), and *enterohemorrhagic* (bloody diarrhea).

Although all infections are serious, the most critical of the coliform bacteria is a type called *E. coli* O157:H7 (enterohemorrhagic) which can cause hemorrhagic colitis and renal failure in children. It is a life threatening form of *E. coli* which must be carefully avoided and can be controlled through good hygiene and thorough cooking of meat products. A major U.S. food poisoning outbreak due to this particular microorganism has recently occurred. This incident and its disastrous results are described in Chapter 9. Another type, O27:H2O, exerts its effect through toxicoinfection and has caused disease outbreaks through the consumption of raw milk soft cheeses such as Brie and Camembert.

Because *E. coli* is a normal inhabitant of the gastrointestinal tract, it was always considered harmless to health in the past, and only used as an indicator of the sanitary conditions of food and water. Since the full impact of *E. coli* foodborne infections has only come to light recently, much greater attention is currently being paid to this microorganism. As humans are the chief carriers of this bacteria, contact with infected food handlers is a major source of the infection. Animals are also a reservoir, and a great deal of care in the production of meat and dairy products is required. Again, even vegetables fertilized with unsterilized sewage or manure can be vectors in the transmission of this disease.

Vibrio *infections*

Vibrio bacteria are responsible for some rather infamous diseases. Cholera is caused by *Vibrio cholerae*. This disease has resulted in worldwide epidemics (pandemics) that continue to flare up. The symptoms are characterized by vomiting, severe diarrhea, thirst, cramps, weakness and, if not properly treated, death. The tremendous loss of body fluid that accompanies the diarrhea not only results in critical dehydration, but also serves to spread the bacteria in the sewage. Cholera is therefore much more prevalent in developing countries, where the lack of water treatment allows continuing cycles of reinfection to occur. Although water is the most important factor, foods also bear a heavy responsibility. More recently, the consumption of shellfish or raw fish, grown in cholera-containing water, has proven to be a vector in the spread of this disease.

Another serious type of *Vibrio* is *Vibrio parahaemolyticus*. It is similar to *V. cholera* except that it requires a marine environment. It is therefore a problem with seafood. It is particularly troublesome where fish are eaten raw, as in Japan and Korea. Symptoms are similar to other gastric toxicoinfections. There have been many deaths due to this bacteria, particularly among older people, but the symptoms are generally not too severe and most victims recover in a few days.

Vibrio vulnificus is the last of these troublesome bacteria. It is responsible for a condition called fulminating or explosive septicemia, which is often a fatal disease. Again, it is associated with the consumption of raw shellfish, such as oysters and clams.

Other foodborne bacterial diseases

Some other fairly common bacteria-related intoxications and infections are due to *Yersinia enterocolitica, Bacillus cereus, Aeromonas* spp. and *Streptococcus* spp. Although not as well-publicized as the other foodborne bacterial diseases, they are nevertheless quite serious, and responsible for some pretty deplorable medical conditions including diarrhea, septicemia, ileitis and meningitis.

Mold Diseases

Aflatoxin poisoning, ergotism and aleukia

Food intoxications can also result from various molds. The poisons are called *mycotoxins*, the most familiar being *aflatoxin*, produced by the mold *Aspergillus flavus*. Other toxins are produced by the molds *Fusarium* and *Claviceps* which contaminate grains such as rye and wheat. As an example, *Claviceps* infects rye and its consumption causes the disease *ergotism*. This disease, St. Anthony's fire, is characterized by muscle inflammation, excruciating pain, gangrene and ultimately death. In centuries past, ergotism reached epidemic proportions. *Fusarium* mold contamination of grain was more recently responsible for an epidemic of toxic aleukia in Russia during World War II. Thousands of people died a tragic death as a result.

Molds usually grow on damp grain or oilseeds (peanuts) where they excrete the toxin. Unfortunately, most mycotoxins are very resistant to heat and therefore cooking has little effect on their destruction. The best course of action is simply prevention of contamination and growth through the proper harvesting and drying procedures. This limits the ability of molds to grow and produce mycotoxins. Products such as peanut butter are subject to uniform aflatoxin contamination simply because the peanuts are homogenized. If two peanuts out of a large batch are heavily contaminated, the toxin gets diluted and distributed throughout the product. That is why peanut butter is, generally speaking, very carefully controlled by both manufacturers and government laboratories. On the other hand, because it is not easy to spot all the moldy specimens in a batch of eating nuts, it is important that consumers make sure they don't consume dark, mottled or moldy ones. If it tastes bad, don't swallow it—spit it out! The old expression "A little mold is good for you," is nothing more than bad advice.

Viruses

Hepatitis, viral gastroenteritis and poliomyelitis

Viruses are also agents of foodborne disease. Hepatitis A, Norwalk gastroenteritis and poliomyelitis are examples of these diseases. The polio epidemics which broke out during and after World War I, as well as more recent outbreaks, were associated with the consumption of unpasteurized milk. In fact, any food prepared by an infected food handler can transmit the disease. The same can be said about viral gastroenteritis. Since it is considered that most outbreaks result from the handling of foods by infected workers, strict obedience of sanitary hygiene regulations and adherence to recommended food manufacturing practices are the most effective means of eliminating these threats.

One of the more common of the foodborne viral pathogens is the Norwalk virus. In the U.K., it is known as the SRSV (small, round, structured virus). The name Norwalk simply derives from a small town in Ohio where an outbreak attributed to the virus occurred. The symptoms are usually vomiting, nausea and diarrhea.

Shellfish raised in waters contaminated with human waste have been implicated in the transmission of hepatitis. The consumption of these shellfish in a raw state is now considered to be an unacceptable health hazard in some parts of the United States [67]. A massive outbreak occurred in Shanghai recently, where more than 16,000 people were estimated to have contracted hepatitis as a result of eating raw shellfish [68]. The ensuing person-to-person transmission of the virus from the original victims resulted in a total estimate of over 300,000 cases of hepatitis. Although it is usually not a fatal disease, hepatitis can be very debilitating, and its symptoms may last for many months.

Bovine spongiform encephalopathy

Bovine spongiform encephalitis, or BSE, is a recently recognized disease of cattle. It manifests itself as a brain disorder and after a period, death ensues. The precise cause of BSE is not known at this time. The disease agents are not recognized as bacteria, and they do not behave like typical viruses. There is speculation that the disease may have resulted from a similar condition in sheep (called scrapie) through the incorporation of rendered offal in cattle feed [69]. As a result, the traditional rendering procedures have recently been modified to eliminate this possibility.

There is a rare condition called Creutzfeldt-Jacob disease which is a similar ailment in man, but there is no evidence that it can be transmitted via food. In fact, there is a similar, if not identical disease in Papua New Guinea called kuru. It has been conjectured that cannibalism, and particularly the habit of eating a victim's brain, was the means of transmission. This is an interesting view for tabloid newspapers, but the available information is simply not enough to be scientifically accepted at this time.

Parasites

The last group of organisms associated with foodborne diseases are the parasites. Parasites really cannot be classified as microorganisms since tapeworms, roundworms and flukes can be very large. Although foodborne parasitic infections are not nearly as great a problem in developed countries as they are in developing countries, they are still considered to be extremely serious on a global scale.

The smallest of the parasites are the *Protozoa*, which are single-celled organisms. The three most important ones in foods are *Entameba histolytica*, *Giardia lamblia (intestinalis)* and *Toxoplasma gondii*.

Entamebiasis

Entameba histolytica is the cause of amebic dysentery (bloody diarrhea), which can be fatal if the parasite migrates to the liver, lungs or brain. It is fairly common where there is poor sanitation and inadequate water treatment, such as in most developing countries, but it has also been known to cause outbreaks in developed countries. An example was the Chicago amebiasis epidemic of 1933, which resulted from the accidental contamination of drinking water with untreated sewage [70]. Chlorine treatment alone is not sufficient to kill this parasite in drinking water. In all areas where it is found, filtration and boiling are required. *Entameba* can also be transmitted via foods when the food handlers fail to observe good hygiene. Raw vegetables are particularly potent carriers of this disease, if night soil (unsterilized human feces) is used as a natural fertilizer.

Giardiasis

Giardia lamblia (intestinalis) is a Protozoa which has a few flagellae or tails on its body in order to guide its locomotion. This organism is found throughout nature and, when ingested, causes an infection of the small intestine known as giardiasis, and results in diarrhea, cramps and weight loss. It is the most commonly reported protozoan parasite in the world today, and occurs in both developed and developing countries [71]. Unfortunately, *giardiasis* can persist for many weeks or even months, if it is not diagnosed and treated properly. Its spread is very similar to that of *entamebiasis*, and therefore the preventative measures required are also similar.

Toxoplasmosis

Toxoplasmosis is a disease caused by a tiny Protozoa which requires cats to ensure its spread, since felines are a necessary part of the organism's life cycle. (If you feel obliged to kiss your cat, please do it from a distance.) *Toxoplasma gondii* is quite common all over the world, and the incidence of prior infection (as determined by the presence of antibodies in blood samples) in the United States has been estimated to be as high as 50% of the population [72]. This disease is serious because it can attack nervous and muscular tissue. Cysts of this parasite have been found in pork and lamb, and ingestion of undercooked meat products from these animals is a potential cause of *toxoplasmosis*. These cysts are, however, very susceptible to treatment with low levels of ionizing radiation.

Cryptosporidiosis

Cryptosporidium is the unusual name of a small protozoan parasite that is steadily becoming a greater concern in the food industry. The parasite causes severe human diarrhea which can be fatal under certain conditions. As yet, there are no drugs or drug combinations developed to which *Cryptosporidium* responds. Consequently, both human and animal hosts of this parasite must allow the disease to run its course, and the only possible therapy is intravenous rehydration. The disease may last for several weeks and bowel movement frequency can be as high as twenty-five times per day. Although in some people the symptoms may be similar to traveler's diarrhea, in immunocompromised people such as AIDS victims, it is a life threatening disease. Recent reports indicate that it is one of the most prevalent of parasites [73]. Although pasteurized, canned, dried and frozen foods do not appear to be good vehicles for *Cryptosporidium*, there are no particular preventative measures in place today to handle this parasite in fresh foods [74].

Roundworms

The parasitic roundworms include *Ascaris lumbricoides*, *Trichuris trichiura*, *Trichinella spiralis* and *Anisakis simplex*.

Ascaris

Ascaris is a common, rather large roundworm found in humans and other animals. Although they can be 15 – 30 cm long, their actual spread is through the ingestion of their tiny eggs. The eggs are shed in human waste. Again, the use of unsterilized night soil as a fertilizer is a definite cause of reinfection. The worm is found in other animals which can serve as another source of infection. (I even found one wrapped around the yolk of a fresh egg while I was at university.) Symptoms are abdominal pain, vomiting, nausea and a protruding abdomen. It is estimated that over 600 million people worldwide are infected with *Ascaris*.

Trichuria

Trichuris trichiura (whipworm) is quite similar to *Ascaris* and is more common in Europe.

Trichinella

Trichinella spiralis causes the disease known as trichinosis. It is most commonly transmitted through the ingestion of undercooked pork where it resides as a cyst in the muscle tissue. The cyst covering is dissolved in the digestive process, and the liberated larvae mature into worms in the intestine. The adults penetrate the wall of the intestine to produce eggs which hatch, and the larvae move through the circulatory system into the muscles. The symptoms can include painful and inflamed muscles, including those that control breathing and speech, as well as fever. Eggs and larvae are shed in the feces and thus allow the whole process to cycle over again.

Anisakids

Anisakids are roundworms that are parasites of marine fish. The chief source of infection is the consumption of raw fish, and therefore there

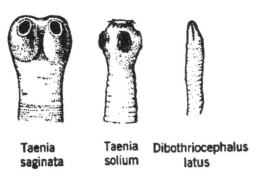

Taenia
saginata

Taenia
solium

Dibothriocephalus
latus

FIGURE 3.4 The heads of human tapeworms.

is a greater incidence of disease in countries such as Japan and Korea. Raw herring consumed in the Netherlands and Scandinavia are also sources of contamination. If the fish are not to be eventually cooked, then ionizing radiation is the only effective way to remove the threat of this parasite.

Flatworms

Tapeworms

Although most people have heard of tapeworms, few people know what they are all about. While they are less common in developed countries than roundworms, these flatworms make their frequent occurrence in developing countries an important problem globally. *Taenia saginata, Taenia solium,* and *Diphyllobothrium latum* are the beef, pork and fish tapeworms of humans. (Pork tapeworm infections are also referred to as cysticercosis.) They are acquired through the ingestion of larval cysts in meat and fish. Where raw or partially cooked beef or pork is consumed, as in certain sausages for instance, the incidence of tapeworm infection can be very high. The consumption of raw fish, or even the tasting of fish soup before it is fully cooked, has resulted in a wide spread of fish tapeworms.

Both beef and pork tapeworms act in much the same way – they cause abdominal discomfort, nervousness and, in some cases, loss of weight. This is understandable, since the worms eat the food that was destined for their host, the consumer. Some tapeworms can span the entire length of the intestine. Diphyllobothriasis is, generally speaking, not too serious but is usually long-lasting. In all cases, proper sanitation and inspection combined with sufficient cooking are a basis of prevention. These parasites are also fully susceptible to low levels of ionizing radiation.

Flukes

Flukes are a type of flatworm called trematodes. They are responsible for a wide number of disease conditions in developing countries. *Opisthorchis viverrini*, is a liver fluke that infects millions of people in Asia. The parasite goes through a complicated life cycle involving one stage in a host freshwater snail and then a free-swimming stage. It eventually is consumed by freshwater fish, whose flesh it promptly invades. If the fish are not properly cooked before eating, the parasite infects consumers. Its effects are serious, and include liver obstruction, fibrosis and jaundice. It has been estimated that 6 million people in Thailand alone suffer from this infection.

Other foodborne trematode conditions include the sheep liver fluke disease, which affects meat consumers in Latin America and the Caribbean, and the lung fluke disease which is serious in many tropical countries and results from eating infected freshwater crabs and crayfish.

The Costs of Foodborne Disease

Foodborne diseases and the organisms that cause them are not a pleasant topic. We have a tendency to visualize them in all our foods, and imagine the very worst of all the people involved in the food manufacturing and handling business. The most positive consolation is that most people wash their hands quite carefully and very regularly for a long time after they have become acutely aware of the issue.

As unpleasant as the subject is, it is one that we must be cognizant of. Foodborne diseases are common, and are far more important than most people imagine. Preliminary appraisals of the cost of foodborne diseases indicate that their economic impact is very significant [75–77]. When factors such as lost labor or income, medical and hospitalization costs, legal fees, and other associated costs are taken into account the estimates run into billions of dollars. The most recent estimates for the United States, which were based on 1985 figures, places the total costs resulting from intestinal infectious diseases above $25 billion annually. This includes medical costs, as well as the cost of lost productivity. Even though $25 billion seems like an enormous sum, it is probably only the tip of the iceberg, since the value of lost opportunities, ruined reputations, and grief due to sickness and death are not possible to calculate with any real meaning, particularly for the victims.

Although it may not provide any comfort, the fact that we do not suffer more from foodborne diseases than we do is a tribute to the professionals in the food and health industries who take their jobs seriously. In the supposedly good old days, people routinely died in great numbers from these diseases. Today, in countries where the professionals and control measures they create are not available, great numbers of people continue to die. Foodborne diseases are a very serious business. They are a business to be left to trained professionals, and not to amateurs, even if they happen to be well-meaning. It is certainly not a subject to be left to nonprofessionals who downplay the issue in order to promote their own parochial interests.

CHAPTER 4

The Use of Irradiation to Prevent the Spread of Foodborne Diseases

Why Use Food Irradiation?

Considerable attention has been paid by most governments to ensuring that consumers have information on how to minimize the possibilities of foodborne disease occurring in the home [78]. However, the ubiquity of disease-causing agents in our natural environment makes it nearly impossible to fully prevent food contamination. To compound this problem, without the proper measures in place to thwart the spread of food-related infection, the cycle of reinfection never ceases. We therefore add to the ever-increasing risk simply by refusing to take active measures to eliminate pathogens from our foods. The prevailing growth rate in foodborne diseases is simply a reflection of this situation. There isn't the slightest reason to doubt that this predicament will continue to worsen until we consciously and deliberately decide to break this cycle. If the figures on the impact of foodborne diseases appear shocking now, just try and project what they will be in ten or twenty years.

Considering the problem from a lofty and virtually meaningless viewpoint does little to resolve it. Almost everyone would like to turn back the clock to a time when the world was less populated, and there were less pressures upon the limited resources of our planet, but it is simply not possible. We must therefore make the wisest accommodations for our increasing needs. In the food system, this has meant more intensive production of plants and animals. Larger production units and greater concentrations of livestock are more efficient, but they also increase the potential exposure to contagious pathogens and promote

71

their spread. Our changing lifestyles have resulted in much more of our food being prepared by others in large-scale establishments, as well as the consumption of many of our meals outside the home. This system of mass production requires sophisticated distribution systems with very strict supervision. The more complex the system gets, the greater the chance of breakdowns, and the greater the need to employ the most effective control measures.

The use of food irradiation has been shown to be extremely effective in decreasing our exposure and risk to offending foodborne pathogens. It has been used to reduce or eliminate various pathogens in beef, poultry, pork, lamb, fish and seafood [79−83,187−195]. It has also been very successful in eliminating or greatly reducing the heavy load of offending pathogens in dried vegetables, herbs and spices. Although these latter commodities, because of their low moisture content, do not support the active growth of microorganisms, they can cause major problems when they are added to other foods, such as ground meat mixtures which are used in hot dogs or sausages. Ionizing radiation has also been used to eliminate pathogens from certain dairy products, and from naturally fermented products such as Chinese soy sauce [84]. It has also been found to be useful in the new generation of modified-atmosphere packaged products, which were designed to have a longer shelf life under normal refrigerated conditions [85,196−203].

Depending upon the product and its place in the distribution system, the methodology used to treat foods will vary. For instance, if the goal is to provide trichinosis-safe pork, the most convenient method would be to simply irradiate a whole side of the animal. This can be accomplished easily with the same sort of conveying equipment that one sees in a standard packing plant. Once it has been irradiated, *Trichinella spiralis* cannot reinfect the pork after it has been slaughtered, and the side can then be transported to the butcher shop or the supermarket without any concern for the offending parasite. The same process can be used to make beef, veal or lamb safe from the microscopic tapeworm eggs that may have been released from the intestines during the slaughtering process. This process obviously does not remove the necessity for continuing to adhere to the most hygienic practices. Routine inspection to ensure that the best possible handling and manufacturing practices are employed will always be required. Irradiation will be employed where even the best production procedures are still no guarantee against spoilage or disease.

This circumstance is identical to that of milk almost a century ago. As previously mentioned, certified milk was guaranteed to have been produced following the best possible procedures of cleanliness. Even the medical authorities endorsed certified milk. But everyone knew that the procedures employed were no guarantee against diseases from pathogenic bacteria in the milk. The only guarantee was to destroy the bacteria. It is no different with other products. Good manufacturing practices (GMPs) are no guarantee against pathogens. GMPs exist to give consumers a high-quality product, and to bring the risk of foodborne disease to the minimum achievable with a given technology. The only guarantee against disease and spoilage is to terminate the causative agents. And even though the pasteurization process was found to be so effective against bacteria, modern procedures for the distribution and processing of milk are far more stringent than they ever were in the past. So much for the old complaints that pasteurization would discourage good manufacturing practices.

In the case of most parasites, their complicated and specific life cycles do not normally allow them to live freely in the environment, and infect food by simple, haphazard exposure. The possibilities of other parasites reinfecting a food product once it has been treated are very remote, if not impossible. Therefore the dual advantage of killing the parasites before they get to the consumer is that it prevents infection from taking place and, just as importantly, it prevents continuation of the reinfection cycle. This may eventually lead to the total elimination of certain parasites from the food system, and may also remove the need to irradiate particular foods in the future. Perhaps this is too optimistic a view, but at least irradiation treatment will stop the problem of reinfection due to consumption of contaminated products. To most people, the very idea of walking around with parasites is pretty revolting. When most individuals sit down to eat food, they want to be certain that they are doing the eating—not the other way around!

Unfortunately, the problem is not quite as simple with foodborne pathogenic bacteria. These microorganisms are very common in the environment, and can start up the process of infection simply by being exposed to the right food. Because of this, measures are required to ensure that irradiated foods are not reinfected in any way prior to reaching the consumer. This usually means that the foods are packaged first, and then irradiated. Once properly packaged, the exclusion of bacteria from the environment, or from workers who are ill or have poor hygienic habits, can be ensured. Thus, products such as poultry or seafood would first be packaged and then irradiated. Packaged frozen products can also be effectively irradiated. In this way, the product will reach the consumer without the possibility of external contamination occurring after treatment.

Foods that are spoiled cannot be improved through irradiation. It is important to understand this. In the same way that sour milk cannot be unsoured by pasteurization, decayed food cannot be miraculously made wholesome through the use of irradiation. If food has been spoiled and contains pathogenic bacteria, then food irradiation will definitely reduce the chances of illness from the pathogens, but the food will still be spoiled. Spoilage is irreversible. When you roast a rotten chicken, all you are left with is a rotten roasted chicken—it is the same with any process. If you put spoiled tuna into a can, you get canned, spoiled tuna, and a spoiled reputation to boot.

Another good example of the application of food irradiation is the processing of seafood. Some of the most beautiful and delicious seafoods in the world come from Southeast Asia. Unfortunately, the distribution of products from the fishing boats to the packing plants, where freezing or chilling is carried out, takes some time. The normal bacteria that these products contain continue to multiply after harvesting, and can negatively affect the quality of the finished products. For instance, it is not uncommon to see some shrimp with dark shell tips. This is due to normal, nonpathogenic bacterial action. Although the shrimp may not be spoiled at this point, unless corrective action is quickly taken, the distribution life or shelf life of these shrimp will be severely limited. By irradiating them, either fresh or frozen, their high quality is maintained through the cessation of spoilage activities [86]. The useful marketing life of the shrimp is extended and they thus maintain their value to both consumers and producers. The same situation can apply to fish [87]. Spoilage is a continuous process and is determined rather subjectively. What is spoiled to some people would be considered perfectly fine by others. Food industry standards are rather rigid in order to ensure that the greatest part of the public would be satisfied with the commercial market quality. That is why it is so important to arrest or delay the spoilage process as early as possible in the continuous chain from harvest to the consumer.

As previously mentioned, shellfish are carriers of many different pathogens, such as *Vibrio cholera*, *Vibrio parahaemolyticus*, *Salmonella* and *Hepatitis A* virus. Since a very large proportion of shellfish are consumed raw, they can expose consumers to great risks. In fact, in some areas, shellfish have for a long time been considered an unacceptable health hazard. This has not reduced their consumption, however. Technologies to reduce this problem have been tried, but even the traditional method of long flushing and depuration (self-purging) of shellfish in clean water has not proven effective in eliminating the danger of viruses as an Australian report has recently indicated [88]. The most current research has shown that food irradiation can either eliminate or reduce the number of these pathogens, including *Hepatitis A* virus, to a level where the risk of shellfish consumption is remarkably reduced [89]. Consumers would still be able to consume the shellfish raw and fresh since the irradiation treatment does not kill the shellfish, nor significantly affect the shelf life. In other words, the irradiated shellfish would be the same in every way as untreated shellfish, except that the risk of hepatitis, cholera and other diseases would be minimized.

There are very many such examples where irradiation can effectively retard spoilage, and result in food products which are safer to consume. If these products look, taste and remain fresh for a longer period of time, it is simply because they *are* fresh for longer. Freshness is far more a criteria of quality than of time. Packaged sliced white bread is fresher after three days than unpackaged hearth bread is after one day. A nice chunk of fresh parmesan cheese can easily be two or three years old. Fresh apples could well have been kept for six months or more in controlled atmosphere storage before reaching the consumer.

The difficulties of preventing pathogens and spoilage bacteria from entering the food system can be tremendous. It often is difficult for consumers to picture the chain of events, and the processes involved in getting food to the table. This is one of the great obstacles in getting people to understand why certain decisions are made in the food industry. Very often, the criticisms made of decisions in the food industry result from a lack of knowledge of the particular systems employed, and the problems they entail. It is beyond the scope of this book to look at a wide range of food processes, but it will be useful to examine one of them. Since poultry is one of the most popular foods we enjoy, this is the industry of greatest interest.

The Poultry Industry

The modern production of poultry starts at the farm or chicken house with either eggs or chicks, both of which may be contaminated. Even if a producer would start with the more expensive *Salmonella*-free eggs, the basic problem would still remain. The young birds are not raised in a sterile laboratory environment. *Salmonella* bacteria can be present in the feed rations, in the pens and in the litter. *Salmonella* can even enter through the help of insects. Once present, the bacteria can quickly and easily spread through the communal water trough. The producer is not aware of the presence of *Salmonella* because the birds do not appear sick. There are only two specific varieties of *Salmonella* that are actually pathogenic to chickens, while many others can affect humans. As a result, the bacteria are almost impossible to eliminate. It would be the equivalent of eliminating all bacteria from the barnyard, or the chicken house. The only possibility of *Salmonella* exclusion would be through the sterilization or irradiation of feed, the full sterilization of pens and litter, the complete exclusion of insects and the use of sterile filtered air—in other words, a sterile laboratory environment. This is neither possible nor practical. With great care and attention, *Salmonella* can be minimized somewhat, but it cannot be eliminated.

Once raised to maturity the load of chickens, some of which are already contaminated with *Salmonella*, are then shipped from the farm to the processor. The proximity of the chickens to each other during transport permits a considerable amount of cross-contamination to take place. Dust, feathers, and litter residue all contain bacteria and are obvious sources of *Salmonella* cross-contamination.

Once the birds get to the processing plant, there are many opportunities for *Salmonella* to survive and spread the contamination further. It has long been known that *Salmonella* bacteria have the particular ability to securely attach themselves to the skin of chickens [90]. When the opportunity arises, this attachment happens quickly and firmly. In fact, this attachment is so strong, that tests have demonstrated that even after forty consecutive carcass rinses, the level of *Salmonella* on the poultry skin hardly decreases at all [91].

After the birds have been slaughtered, they are quickly immersed in a scalding tank in order to make the removal of feathers easier. The ideal temperature for scalding is 50−60°C, but this permits *Salmonella* to survive. Some of the bacteria are thus released into the scalding water, and are free to contaminate other birds. Right after scalding, the chickens are transported to the defeathering machine, which is simply a device with rubber fingers on a revolving cylinder. It isn't difficult to understand why this is another source of cross-contamination because the fingers can transfer bacteria from the skin and feathers of one chicken to the skin of another.

The birds are then eviscerated. This process often releases some of the intestinal contents onto the skin or meat of the bird, thereby increasing the degree of contamination. After evisceration, the carcasses are spray-washed in order to clean them, but as mentioned above, this hardly removes any *Salmonella* at all. The birds are then immersed in chill tanks to quickly lower the carcass temperature and retard the spoilage process. Since *Salmonella* easily tolerate the temperatures employed, this chilling step is a final stage where cross-contamination occurs in the plant. The cross-contamination that results from this final step is almost totally eliminated in certain European operations which employ a chilled spray process rather than the common chill tank. However, there continues to be some reluctance in adopting the spray-chilling process in North America, perhaps because the immersion process allows the carcass to absorb a significant amount of water which ultimately contributes to the finished selling weight of the poultry.

In a sense the poultry processor unintentionally acts like a giant mixer to dilute and spread out the contamination among the greatest number of birds. The number of *Salmonella*-positive chickens leaving a plant is invariably greater than the number going in. As a result, up to 80% of the retail chickens in the U.K. [92] and up to 71% of processed carcasses in the U.S. [93] have been found to be *Salmonella*-positive. Rates of *Salmonella*-contaminated poultry in the range of 15−70% are common in most other countries.

The process of mass-producing chickens in order to bring their price down to a more affordable level is not the original cause of contamination, it merely serves to spread the contamination among a greater number of birds. *Salmonella* is common in the environment, and can therefore easily infect chickens at high levels without making them sick. The mass-processing or for that matter any processing of chickens does not employ a stage to exterminate bacteria. Obviously, chickens cannot be pasteurized in the traditional sense. Until a step for the destruction or significant reduction of bacteria is incorporated into the process, however, there is no way to avoid *Salmonella*-, *Campylobacter*-, and *Listeria*-contaminated chickens on your supermarket shelves. Based on all the evidence and all the tests, scientists and health professionals alike feel that the irradiation of poultry meat is the most effective method of significantly reducing the public health problem of poultryborne salmonellosis [94]. The same can be said for other poultry pathogens, such as *Listeria monocytogenes* [204]. The latest results prove beyond a doubt the tremendous value of irradiating poultry for the consumer and restaurant/catering trade. Very large-scale tests were carried out at kitchen level to determine the effects of both irradiated poultry and spices on the overall hygienic quality of the finished foods. For the first time, the potential for removal of the majority of pathogens from the kitchen can be envisioned [95]. Furthermore, irradiation of chicken makes *Salmonella* much more sensitive to the effects of heat—a critical consideration for those who like their meat and poultry on the rare side [96].

A very similar situation exists with pork, beef, lamb and any other fresh or frozen meat product. The same can be said for all types of fish and seafood. Unless there is a procedure in place to destroy pathogens in these foods, we are no better off now than we were with milk 100 years ago!

Various estimates have been made on the cost of irradiation and it varies from about 1 – 3 cents U.S. per pound of product. At the current price of chicken, this represents a very small percentage of the overall product cost. The decision on whether to pay the slight premium in order to have a pathogen-reduced product should belong to the consumer and no one else, including the retailer. It is up to the food system to allow the consumer that choice. When, and if, these additional costs are balanced against the costs of foodborne diseases, it is clear that the benefits are grossly in favor of food irradiation [97].

Spices

Herbs and spices provide another good example of the application of food irradiation. These products are all grown and dried in the open air. Even when they are covered with fine screens, insects, birds and rodents can deposit their excreta directly onto these products. Herbs and spices can also be contaminated by fine wind-blown dust which contains the very same contaminants. This happens in even the most hygienic operations. Once the bacteria are on the fresh herbs or spices, they cling tightly and are not easily washed off. Once dried, herbs and spices are not a good medium for the growth of microbes. However, when these products are incorporated into high-moisture foods, such as sauces or prepared meats, the bacteria, including the offending pathogens, rapidly flourish and severely contaminate the final product. As seen earlier, even cooking the final product is no guarantee against the bacterial toxins which could have developed after the spices or herbs were added. The most effective way to minimize the hazard of such pathogenic microorganisms is to destroy them while they are on the herbs or spices.

The components that give the flavor to herbs and spices are called essential oils. Most consumers are familiar with or may have heard of garlic, pepper or clove oil. Heat or steam treatment in various forms has been recommended as a means of destroying bacteria on herbs and spices, but the essential oils are generally heat sensitive. Therefore, the most practical way to treat these materials is with a cold process. In the past, toxic gasses, such as ethylene oxide, were used to eliminate microorganisms on herbs and spices. Ethylene oxide, in particular, had some critical problems. Aside from being a toxic gas, ethylene oxide forms a flammable and explosive mixture with air, and can thus be a hazard to workers. Nevertheless, it was the most popular treatment for the elimination of bacteria in spices for many years. Most consumers are not aware of this, because there was never any indication of this treatment on food labels, even though there was a possibility of residues remaining on the spices. As a result of these problems, the use of chemical fumigants has either been banned or severely restricted in many countries. Since the dilemma of pathogens in herbs and spices still remains, scientists consider food irradiation to be the most effective and safe method for their control [98,205,206].

There are countless other examples which can be cited where food irradiation has been shown to be the most effective and safe means of reducing the threat of offensive pathogenic organisms in our foods, but it is time to move on to other issues.

In summary, we must be aware of the realities of food production. The natural environment contains organisms that can harm us. Despite all the precautions that are taken to protect our foods from these pathogens, it is virtually impossible to guarantee their total elimination by hygienic practices alone. All our food plants and animals cannot be grown in sterile or artificial environments. Where offending microorganisms cannot be removed, they must be destroyed or rendered nonoffensive. These problems will not be solved by listening to nostalgic or sanctimonious pontifications about the way life should ideally be. These problems are scientifically well-defined and require the most beneficial technical solutions.

We have become acutely aware of the problem of foodborne disease.

It is not a new problem. We are more aware of it now because our growing population has dictated the need to move towards a system of mass-produced foods for greater efficiency. As a result, if contaminated, foods have the potential for making a very large number of people sick in one single outbreak. It is therefore an issue that must be addressed practically and promptly. People are getting sick and dying at this very moment. This is the time when the problems have to be solved—purposeless delays can no longer be tolerated.

We cannot turn back the clock. One quarter of a century ago, the purveyors of doom said we would not be able to feed our planet by this time. If the world had heeded their call, we would, indeed, be unable to feed our population today. Yet, here we are producing more food of a higher quality than we could have ever dreamed. This is not the result of armchair philosophers sitting with their feet up on a desk and advising us to return to the good old days. This is because of the ability of our professional scientists and technicians to innovate and face problems with real, and not theoretical, solutions. Because they are real solutions, they are seldom perfect. They must be continually improved. However, the judgments made on areas for improvement must be based on objective information, not on phobias or biases.

Food irradiation has been proven to be a very powerful and practical means to reduce the pathogenic potential of solid foods, just as pasteurization has been to liquid foods. As our short history above has shown, scientific information and, above all, common sense eventually permitted the benefits of milk pasteurization to reach consumers, despite the countless illnesses and deaths which could have been prevented had the method been employed earlier. This same common sense will eventually prevail in the case of food irradiation. The question is, how long will it take, and how many people will have to die before this happens?

CHAPTER 5

The Prevention of Food Losses after Harvesting

The Extent of Losses

By the year 2000, it is projected that the world population will approach 7 billion people. This level of population will add to the pressures upon our limited resources, and significantly affect our ability to feed ourselves. In order to cope with increasing food demands, the two traditional lines of action—that is, reduced population growth and increased agricultural production—will have to be markedly supplemented with the parallel activity of reducing all possible food losses both during and after harvest. Although prevention of these losses is generally not a priority consideration for most of us, it is an activity whose importance cannot be overemphasized.

In developing countries in particular, enormous losses can result from spillage, contamination, insects, birds, rodents and normal biological spoilage during storage. Reliable studies indicate that such losses amount to tens of millions of tons per year, valued at billions of dollars. This is not particularly surprising, when one considers the losses which can occur in well-organized food systems. For example, the International Rice Research Institute estimated that for postharvest rice operations in the Philippines, there is a 1−3% loss for harvesting, a 3−7% loss for handling, a 2−6% loss for threshing, a 1−5% loss for drying, a 2−6% loss on storage, and a 2−10% loss on milling. Therefore, at the lowest end we expect an 11% loss, and a 37% loss at the upper end, even when fairly advanced agricultural techniques are used.

In semiarid Africa, the situation is far worse. Storage losses alone

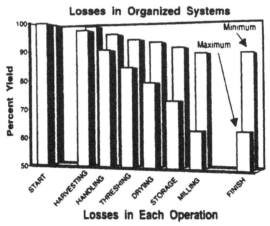

FIGURE 5.1 Example of food losses.

have been estimated from a low of 10% in cereal grains to a high of 75% in the more vulnerable pulses (peas and beans). With some crops, significant insect damage can occur even before storage, and resultant losses may run up to 60%. If these levels of loss are multiplied by the actual production figures, the enormity of the problem readily becomes apparent. These figures do not deal with hypothetical new miracle crops, which may have significantly increased yields, or the ability to grow on poor soils. Neither do they refer to the wonderful new varieties eventually expected from agricultural biotechnology. We are dealing with actual, ongoing conditions, and crops that are currently being produced. The situation exists whereby the methodology currently available, if applied properly, can result in genuine yield increases between 20 and 40%. The economic and social impact of this can be staggering, particularly since this yield can be directly converted into increased consumption on the part of our fellow humans.

What precisely are we talking about, when we refer to postharvest food losses? First of all, food losses themselves vary greatly and are a function of the crop variety, the pest combinations in the environment, the climate, the system of harvesting, and the type of processing, storage, handling and marketing the products undergo. In addition, the social and cultural environment may also have an important influence.

It may be useful to illustrate some of these figures with a specific example. In just one Indian city, Bombay, it is estimated that well over 4,000 tons of cereals are lost annually due to rodent damage alone. A single ton of cereal will feed six people for an entire year. Bombay is not the fabled Hamelin and has no mythical Pied Piper. But there are practical, technical procedures to control rats. Rodent damage alone, if prevented, can supply 25,000 people in Bombay the equivalent of a year's supply of food. Compound this figure by all the other major elements involved in postharvest losses and one can eventually get a perspective on the effect of this problem on human life.

Causes of Losses

Several factors lead to the cumulative causes of postharvest food losses. These include inefficient harvesting and drying methods, poor processing techniques, inadequate methods of storage and distribution and, even in the home, poor preparation or use of the foodstuffs.

The more traditional methods of grain distribution in some parts of the world make exclusive use of, for example, jute sacks rather than bulk shipments, and are thus more susceptible to leakage, or insect infestation. Traditional marketing systems can contribute to reduced returns to farmers by involving several changes of hands before the produce reaches the consumer. Many methods of storage provide insufficient protection against the elements, and inadequate road and transport systems make distribution to key points extremely difficult, particularly during critical seasons. Even when dealing with newer technologies such as irradiation, which may have a role in one part of the system, it is important to ensure that other key factors in the entire flow of goods to the consumer are fully addressed.

The actual causes can be categorized into two main groups, the primary and secondary causes of postharvest losses.

Primary causes

The primary causes of loss are those that directly affect the food. They may be classified into the following subgroups:

(1) Biological—Biological causes relate to the direct consumption of food by insects, rodents, or birds, resulting in the immediate disappearance of food. Sometimes the level of contamination by excreta, hair and feathers can be so high that the remaining food will be condemned for human consumption. Insects cause direct weight losses to the quantity of food, and quality losses because of the webbing, filth and odors they leave in the foods.

(2) Microbiological—Although microorganisms—fungi and bacteria—usually consume only small amounts of food, they can damage it to the point where it becomes unacceptable because of rotting or other defects. Toxic substances such as aflatoxin, produced by molds, can cause food to be condemned and useless even as a feed material.

(3) Chemicals—Many of the natural chemical constituents present in foods can react with each other, causing loss of color, flavor, texture and nutritional value. An example is the chemical reaction that causes browning in dried fruits.

(4) Biochemical reactions—A number of natural, enzyme-activated reactions can occur in foods during storage, giving rise to off-flavors, discoloration and softening. The simple bruising of bananas is a good example. Another illustration of this problem is the unpleasant flavors that develop from frozen vegetables that have not been blanched. Blanching deactivates the responsible enzymes before freezing.

(5) Mechanical—Bruising, cutting, excessive peeling or trimming of agricultural products results in mechanical losses.

(6) Physical—Excessive or insufficient heat or cold can spoil foods. Improper atmospheric conditions in closely confined storage at times causes very significant losses. In fact, this often happens in home refrigerated storage.

(7) Physiological—Natural respiratory losses which occur in all organisms account for a significant level of weight loss, and also generate damaging heat. Changes which occur during maturation, such as ripening, wilting and sprouting, also increase the susceptibility of the product to mechanical damage or infection by pathogens. A reduction in nutrient level and consumer acceptability can also accompany these changes.

Of the primary causes, the biological, microbiological, mechanical and physiological factors are responsible for the majority of the losses in perishable crops.

Secondary causes

Secondary causes of losses are those that lead to conditions which encourage a primary cause of loss. They are usually the result of inadequate or even nonexistent storage structures, technologies and quality control. Some examples are:

- inadequate harvesting, packaging and handling skills
- lack of appropriate containers for the transport and handling of foods
- inadequate storage and transportation facilities
- most important of all – the lack of knowledgeable management and technology that is essential for maintaining food in good condition during storage and marketing

It is impossible to cover in detail all the implications of inadequate postharvest practices. As an example, however, the issue of insect damage will be examined in order to provide some perspective on the problems. It is fairly evident what the consequences of direct food consumption by insects is – there is simply less food available for people. A very important secondary consequence is the actual removal of food commodities from international trade as a result of the need for quarantine treatment.

The Need for Insect Quarantine

Today's world is characterized by intensive traffic of both people and goods. Modern air, sea and ground transportation bring the remotest places of the earth into close contact with one another. This situation can be very desirable and can also be of great benefit to us. However, it also creates some risks which did not previously exist. Among these hazards is the movement of pests, such as insects, from country to country.

These pests originally evolved through an equilibrium with their natural environment. However, when such pests move from their home country to new areas without their natural enemies, they can flourish and become unmanageable. This has often resulted in serious consequences for the new location. A well-known example of this sort of phenomenon is the African bee (unjustly referred to as the "killer bee"), which was imprudently introduced to Latin America and is now rather difficult to control. Introduction of insects into new environments has become the major avenue by which they effectively become pests.

Insects are easily distributed by international trade and by tourism. In order to prevent or minimize this risk, many countries have established quarantine measures fully supported by laws and regulations. These measures can create significant barriers to international trade and the free flow of plants and plant products, but they are fully justified from the receiving country's point of view.

One of the best examples is the Mediterranean fruit fly, or medfly, which is perhaps the most feared pest of many countries. The medfly originated in East Africa, and spread during the nineteenth and twentieth centuries to many countries in the tropics and subtropics. The female fly first deposits her eggs on the surface of the fruit. Upon hatching, the larvae dig themselves into the flesh of the fruit, resulting in physical damage and microbial decay. All types of fruit, including citrus fruits and many other tropical fruits are among the casualties. Damage from medflies can result in losses of 80–100% of the crop. It is no wonder that countries which produce fruit of any kind don't want to let a single fly enter. If preventing the entry of small flies may appear difficult, it is not hard to imagine how complex it is to prevent tiny medfly eggs from coming in. Countries, therefore, resort to quarantine measures.

Quarantine measures

Because of the disastrous potential consequences of such pests, quarantine measures are enforced by law. Strict regulations prohibit the entrance of plants or plant products (which may hide the quarantined pest) from countries where the pest is known to exist. Inspections are carried out at all ports of entry with the objective of intercepting and destroying contaminated material.

Exemption from such quarantine is granted only on assurance that agreed-upon measures to effectively disinfest the plant material have taken place. Such measures could consist of specific treatment of the material, either at the country of origin, at some point during transit, or at the final port of entry, if suitable conditions exist. Quarantine enforcement officers carry a very great responsibility, which only the importing country's farmers appreciate. If more consumers realized the potential economic consequences to their country, they might refrain from adding to the problem by smuggling in "harmless" fruit or plants from the exotic countries they visit.

Disinfestation of plants and plant products

Quarantine treatments can be placed into three categories.

(1) Chemical treatments – Chemical treatments use fumigants for pests which occur inside, or on the surface of, the plant or plant product.

   Until a few years ago, ethylene dibromide (EDB) was the most common fumigant in general use. In 1984, the U.S. banned its use on grain after it became recognized as a possible carcinogen. Since then, its use as an agricultural pesticide has been banned in most countries. An unfortunate victim of this ban on ethylene dibromide was the Caribbean Basin Initiative. This was a trade development program organized to promote the importation of products (primarily fruit) into the U.S. from the Caribbean region. Before the program got off the ground, the use of EDB was banned, and all treated fruit was excluded from import. Since no other alternative means for disinfestation was available at the time, it spelled disaster for Caribbean exports. They have never recovered from the shock.

   Nonchemical methods comprise the next two categories.

(2) Physical treatments – Physical treatments depend on high or low temperatures applied in various ways. Cold treatment of fruits as a quarantine measure for fruit flies is an old method. It involves a long storage of 10 – 15 days of the packed fruit (citrus) at temperatures ranging from 0.5°C to 2.0°C. Unfortunately, this method is from three to five times more expensive than EDB was. Another critical problem is that there cannot be a single breakdown of cooling or monitoring equipment during the entire period, or the treatment may not work. Cold treatment is also relatively damaging to many varieties of fresh fruits or vegetables, such as papayas, avocados and mangoes, which cannot tolerate such treatment.

   High-temperature treatments, such as repeated hot water dipping and vapor heat, are also used on papaya, for example, but the fruit quality can suffer substantially. The fruit must be picked fairly green in order to survive the treatment, and thus the flavor and aroma simply do not come close to the original fresh fruit.

(3) Ionizing radiation – A great amount of work has been carried out on the use of ionizing radiation for the disinfestation of fresh fruits and

vegetables for quarantine treatment. Recent studies have shown that rather low dosages of ionizing radiation will very effectively control fruit fly [99 – 101], and other insect problems [102,207,208]. This makes the use of irradiation for quarantine treatment a very practical possibility. It does not work equally well for all fruit, but for those that are appropriate, no other method comes close to delivering the top fresh fruit quality that food irradiation does.

In the future, the current decline of pesticide use in agricultural production will also be experienced in quarantine treatment. However, quarantine regulations are here to stay, and quarantine treatments will remain an important part of the plant protection domain.

Irradiation will find a most important application in that domain. But, as mentioned earlier, it has to be part of an overall postharvest management system. Integrated pest management control practices in the field, proper harvesting prior to the most sensitive stages of fruit susceptibility, and careful removal of infested fruit will support and contribute to all efforts made to eliminate this serious problem.

Insect quarantine measures serve as only one example. There are several other applications for food irradiation in the reduction of postharvest losses. Although extensive research has documented the usefulness and the safety of ionizing radiation for these purposes, its potential value can only be realized if it is put to practical use.

The actual dose of radiation employed in any food processing application represents a balance between the amount needed to produce a desired result and the amount the product can tolerate without suffering any unwanted changes. In some particular cases, food irradiation may not be as effective as the current chemical or heat treatments, and should not be recommended [103,209]. In fresh fruits and vegetables, certain levels of irradiation may sometimes cause softening and increases in tissue permeability (watery texture). On the other hand, since irradiation at the right dose level slows down the rate of ripening for fresh fruits and vegetables, properly stored and packaged products remain in usable condition considerably longer than they would without radiation processing. It also allows fruit to be picked in a riper condition so it has a more desirable taste and texture.

Some other examples for the use of radiation to enhance the safety and quality of food are given as follows.

Control of Sprouting and Germination

Low dose irradiation treatment inhibits sprouting of potatoes, yams, onions, garlic, ginger and chestnuts. In the case of potatoes, irradiation has a marked advantage over the current methods of sprout control for the consumer. In order for potatoes to last through a season without spoilage due to sprouting, they are either treated with chemicals or left in cold storage. In the case of cold storage, the low temperatures inhibit sprouting as long as the product remains cold (3.3−4.4°C). Not only is this process very expensive, but as soon as the consumer buys the potatoes, the sprouting process starts all over again, unless they are kept cold. This severely limits a consumer's choices, because the only way to manage sprouting in the home is to buy small packages and keep them refrigerated. This is particularly unfair to poorer consumers who cannot take advantage of bulk purchases, since there is no room in the refrigerator for them. Processors who make frozen french fried potatoes do not particularly care for very low temperature storage because the potatoes don't "cure" properly, and consequently can't develop the preferred golden color when they are fried. Therefore, growers and storage operators often resort to chemical inhibitors.

Chemicals such as maleic hydrazide, isopropyl-*n*-chlorophenyl car-bamate (Chloropham) and 1,2,4,5-tetrachloro-3-nitrobenzene (Tec-nazine) are currently being used on potatoes. They can be applied while the potatoes are still in the field, or they can be dusted onto potatoes as they are put into storage. These chemicals are convenient and effective. The potato processors like them because potatoes can then be stored at higher temperatures and "cure" properly. Unfortunately, these chemi-cal sprout inhibitors can leave residues on the products. In fact, these residues can even build up in the common french fryers that we see in many fast food restaurants [104].

Food irradiation is a safe and remarkably effective sprout inhibitor. It leaves no residues and allows storage at higher temperatures. There is little doubt that it is the most practical method currently available.

Controlling the germination of barley during malting is of considerable economic importance in the brewing industry. Very low doses applied to air-dried barley do not prevent sprouting during malting, but do retard root growth. In this way, high-quality malt can be obtained, while the losses resulting from too rapid sprouting are reduced. Since this effect of radiation processing persists for many months, treatment can be applied before the barley is put into storage, with the added benefit of destroying any insect pests that may be present in the grain. In fact, small irradiation doses (0.01 – 0.10 kGy) can even stimulate the germination of barley, a result that can be used to shorten the malting process and increase the production capacity of malting plants. Low-dose irradiation has also been successfully used to stimulate and control seed germination in order to maximize the development of ascorbic acid and riboflavin in soybean sprouts [210].

Shelf Life Extension of Perishable Foods

One of the principal uses of food irradiation is for killing the harmful microorganisms that cause spoilage or product deterioration. The useful shelf life of many fruits and vegetables, meat, poultry, and fish and other seafoods can be considerably prolonged by treatment with combinations of refrigeration and low doses of irradiation that do not alter flavor or texture. Recent Australian studies on scallops and marine fish reveal a significant increase in shelf life without any significant quality deterioration [105,211]. Most food-spoilage microorganisms are killed at doses well below 5 kGy. Various fresh fruits, including strawberries, mangoes and papayas, have been irradiated, labeled and marketed – with consumers indicating a striking preference for them.

Delaying Ripening and Aging of Fruits and Vegetables

Exposure to a low dose of radiation can slow down the ripening of some fruits and vegetables, thereby extending their shelf lives. This effect of ionizing radiation depends upon the level of treatment and the state of ripeness at the time of treatment. A measurable extension of shelf life may be obtained with some fruits in the treatment range of 0.3 – 1.0 kGy. This procedure will increase the shelf life of mangoes by about one week and that of bananas by up to two weeks. Maturation of mushrooms and asparagus after harvesting can be retarded with doses in the range of 1.0 – 1.5 kGy. In addition, recent work has revealed that irradiation can result in a synchronization of the ripening process [106].

Magnitude of Losses

The extent of actual postharvest losses is difficult to determine accurately. Depending upon the year and the country, it is possible to find individual cases with losses ranging from 0−100%. This high variability is dependent upon a number of conditions. Staple foods such as cereal grains can be stored in good condition for several years, whereas perishable foods, such as fruits and vegetables, spoil quickly unless given particular treatment, such as canning, freezing or irradiation. It is obvious that the longer the food is stored, the more opportunity there is for losses to occur. Generally speaking, perishable crops suffer from higher losses than cereals.

The pattern of losses varies widely from country to country. There is a marked contrast between the sites of major losses in developed countries and those of the developing countries. In a typical developed country, losses can be fairly high during harvesting because the agricultural machinery that is used to harvest the crops leaves a good deal of the production in the field, and mechanically damages some of it. Considerable quantities of foods may be discarded at the point of harvest because they are the wrong size, shape or color. But these are planned losses. In developing countries, harvesting losses are usually lower because most of the crop is hand-picked. The amount of material rejected in the developing countries is less, because food is such a precious commodity, and the expectations of quality and uniformity are generally lower than those of developed countries.

In developed countries, losses are usually small during processing, storage and handling, because of the equipment efficiency, the quality of the storage facilities, and the close control of critical variables by highly knowledgeable product managers. In contrast, in developing countries losses in processing, storage and handling tend to be rather high because of poor facilities, and a frequently inadequate knowledge of methods required to care for foods properly. In developing countries in particular, the tragedy of these enormous losses results in severe economic repercussions in regions that are struggling to escape poverty, and the effect is multiplied by a major loss of nutrients to populations that are often malnourished. Food irradiation has limited use at the present time in most developing countries. However, where there is a properly functioning commodity system, particularly for foods that require complex distribution patterns, such as export crops, irradiation is a definite advantage [107−109,212]. It can even be used where the national distribution infrastructure allows products such as yams or potatoes to be pooled before redistribution [110]. (For example, most of the cold storage capacity in India is devoted to potatoes and onions. If

Table 5.1 National clearances.

Country	Fruit	Spice	Vegetable	Cereal	Roots and Tubers	Meat	Seafood
Argentina	✓	✓	✓		✓		
Bangladesh	✓	✓	✓	✓	✓	✓	✓
Belgium	✓	✓	✓		✓		✓
Brazil	✓	✓	✓	✓	✓	✓	✓
Bulgaria	✓		✓	✓	✓		
Canada		✓	✓	✓	✓		
Chile	✓	✓	✓	✓	✓	✓	✓
China	✓		✓	✓	✓	✓	
Cuba	✓	✓	✓		✓	✓	✓
Croatia Rep.	✓	✓	✓	✓			
Czech Rep.		✓	✓				
Denmark		✓					
Finland		✓					
France	✓	✓	✓	✓		✓	✓
Hungary	✓	✓	✓			✓	
India		✓	✓				✓
Indonesia		✓		✓	✓		
Iran		✓					
Israel	✓	✓	✓	✓	✓	✓	
Italy					✓		
Japan					✓		
S. Korea		✓			✓	✓	✓
Mexico		✓	✓	✓		✓	✓
Netherlands	✓	✓	✓	✓		✓	✓
Norway		✓					
Pakistan		✓			✓		
Philippines		✓			✓		
Poland		✓	✓		✓		
South Africa	✓	✓	✓	✓		✓	✓
Spain					✓		
Syria	✓	✓	✓	✓	✓	✓	✓
Thailand	✓	✓	✓	✓	✓	✓	✓
Ukraine	✓		✓	✓	✓		
United Kingdom	✓	✓	✓	✓	✓	✓	✓
Uruguay					✓		
U.S.A.	✓	✓	✓	✓		✓	✓
Vietnam			✓		✓		✓
Yugoslavia (former)	✓	✓	✓	✓		✓	

92

these were to be irradiated, this cold storage might be more effectively used for the storage of higher value fruits and vegetables.) Unfortunately, the cost of equipment and the need for reliable services dictate that it cannot be considered for use at the basic rural level where much of the postharvest losses take place.

Although it is very difficult to accurately estimate, since reporting procedures are not always in place, attempts have been made to approximate the international impact of spoilage and foodborne diseases. As one might guess, the figures are very high and run into the billions of dollars. Todd has estimated $20 billion [111], but considering the great number of areas that go unreported, the figure could easily be ten times as great.

Table 5.1 gives a listing of all the countries that have technically cleared the use of irradiation for specific applications. (This does not mean that irradiated foods are necessarily available in those countries, because it is up to manufacturers and retailers to supply the market with products.)

As in the case of all other technologies, food irradiation cannot answer every problem. Yet, it definitely has an important role in limiting the appalling loss of health, income and food resources the world is routinely up against. There are some who say that it is a technology which no one wants or needs. They presume to speak for all, and they also refuse to acknowledge the available evidence.

In fact, those who have been very vocal against the use of food irradiation have influenced the food system to the point where it has not yet made irradiated foods generally available as a free consumer choice. Advocates against the introduction of irradiated foods have been successful thus far. Although irradiated foods may not be freely available to the public, microbially contaminated foods are. Anti-food irradiation advocates have deliberately linked the process to the nuclear controversy in order to intimidate politicians, legislators, manufacturers and retailers. Faced with the same scientific facts as everyone else, why is it that certain anti-food irradiation advocates are so set against allowing consumers a free and informed choice of irradiated foods?

What has motivated these people to march in front of food stores with placards to try and terrorize retailers or manufacturers? These are the very same tactics used by those who marched in Manchester in 1933 with signs reading "Pasteurized milk will kill your babies." Why are they against the introduction of a safe and beneficial method of food processing that could significantly reduce the consumer's exposure to disease and spoiled foods? When all the recognized health authorities in the world declare that the process is safe and useful, what are the facts these people have that others do not?

CHAPTER 6

Advocacy Objections to Food Irradiation

There is a general perception that consumers do not want irradiated food products. This groundless view has been expressed by a number of vocal "consumer" advocacy groups and has almost been taken as a fact by the press. The word "consumer" is in quotation marks because there are, in fact, two distinctly different advocacy groups that claim to represent the consumer. The first type is the traditional consumer advocacy group which very clearly has the consumer as their one and only priority. The second group is very different indeed. This latter "New Wave" consumer advocacy group has all sorts of other agendas to follow as shall be seen.

Traditional consumer representatives are genuinely interested in defending the cause of consumers. They do so in two rather different ways. The first method is to represent consumer *views*, while the second is to represent consumer *interests* [112]. Representing consumer views consists of relaying the fears and concerns which some consumers have to policymakers in order to sensitize and influence them on the particular issues in question. This particular aspect of consumer representation is open to much abuse, because if the fears or concerns prove to be irrational, or based upon misinformation, the consumer may end up the ultimate loser. In defending the interests of consumers, the normal course of action is to first gain as much authoritative and objective information on the issues as possible, and *then* inform, advise and ultimately represent the consumer. Consumer representatives seek out and approach the most qualified technical authorities they can for this information, and they in turn convey it to their constituencies in a

95

language which is understandable. Their entire focus is the consumer membership. They have been very successful in ensuring that consumers are informed about the foods they eat, and have made the food industry more responsive to consumers in general. Most traditional consumer groups do not confront the various issues with self-interest or prejudice. They try not to be single-issue advocacy groups, because this seldom leads to a balanced perspective. This, however, is not the style of what may be called "New Wave" advocates.

In the '70s the great futurist, Herman Kahn, wrote of the coming "New Class" of intellectuals who will make their livelihoods from language skills and the manipulation of symbols. He said the New Class would dominate the media, and compel controversial policy issues to be discussed in their own neo-liberal terms [113]. Instead of New Class, let us refer to them as New Wave.

New Wave advocacy groups are very different from the traditional consumer advocates, and so are their techniques. They do not approach their issues in an objective manner, but rather as a moral imperative. They accept only that testimony that suits the object of their advocacy and ignore the rest—even if it is the bulk of authoritative evidence. They cull their bits and pieces of information carefully and disregard all that does not serve their purpose. This is very important to understand, because it clearly confirms that their objective is not the honest or balanced representation of consumers or consumer issues.

The least harmful class of this latter group is simply antiestablishment, Category B New Wavers. They make and reinforce the assumption that the "establishment" and the "people" must exist in an adversarial relationship. They consider government part of the establishment, and it therefore cannot adequately represent the people. Thus, they must take on the role of representing the common people. Their attitude is reminiscent of the Maoist movement of the '60s and since the Vietnam war is long over, other issues must occupy their active temperaments. Despite the fact that their arguments are often distorted, their generic antiestablishment view has made government and industry more sensitive and conscious of the need to think of consumers. Along with traditional consumer advocates, their pressure tactics have benefitted consumers through better labeling and more consumer-oriented products. They have also inspired the industry to provide more detailed information to the consumer.

Single-issue advocacy groups (Category A New Wavers) have proved to be far more harmful. As an example, there are single-issue advocacy groups that espouse causes such as environmental preservation, or nuclear-free societies. While these causes themselves are legitimate, and indeed laudable, they do not give anyone the right to use any measures and any means, fraudulent or not, to support them. In fact, some of the issues are so important that Category A New Wavers have done them a true disservice by focusing on emotional arguments rather than facts. The result has been a distorted perspective and a critical loss of precision in tackling the issues. In the case of food irradiation, almost all the anti-groups have an environmental or antinuclear agenda as their priority. Food quality or safety is not their goal, it is their tool to gain support for their primary agenda. It is a case of the means justifying the ends. This is why their evidence is biased and limited in order to exclude all information which does not serve their cause. It is dishonest tyranny hidden behind a veil of self-righteous motherhood. It is also transparently self-serving.

New Wavers are masterful at manipulating language, and at reinforcing a fear of the unknown. Their objective does not seem to be an informed, but rather a misinformed or partly informed public. Perhaps a better way to describe their goal is a myth-informed public. Another goal is to impart public confidence in a very restricted knowledge of the subject matter, in the hope of achieving credibility with their own limited expertise. This is reinforced by setting up scientists, industry and the establishment as adversaries, thus making scientifically established knowledge and techniques appear to be conflicting, and therefore suspect.

Distorted perspectives result in controversy and it is the maintenance of public controversy that keeps New Wave advocates in business. Shaking the confidence of the public in science and in established institutions serves to justify the need of their presence to protect consumers. Controversy itself very often seems to be the object of New Wave advocacy, and they are very effective at generating debate on their terms. One of the problems is that the public press amplifies the position of the New Wave advocates, because controversy sells newspapers. Another problem is that some well-known traditional consumer advocacy groups have been seduced or hijacked [114] by New Wavers and the attention they command in the press. Kahn predicted that New Class "public interest" groups would be articulate and prepared to debate on any issue, whether or not they were qualified to do so [113]. His prediction has come to pass in this instance.

The New Wavers' focus upon single issues often clouds the true objective of their advocacy. It would seem that a key issue in representing consumers would be the overall safety of their food supply. Certainly the verifiable risk of foodborne disease should be considered seriously in any such judgments. Yet the focus is not the safety of the food supply. The focus is irradiation. It begs to question as to whether New Wavers are using the issue of irradiation to serve consumers or whether they are using consumers to serve the controversial issue of irradiation.

Food Irradiation Complaint List

When analyzing the New Wavers' complaints about the process of food irradiation, it is interesting to note the similarity to early complaints about pasteurization. The bulleted items below refer to the categorized list of pasteurization complaints outlined in Table 2.1.

(1) Food irradiation will be used to mask spoiled food.
   - A 1 – Pasteurization may be used to mask low-quality milk.
   - E 5 – Pasteurization gives rise to a false sense of security.
(2) Food irradiation will discourage strict adherence to good manufacturing practices.
   - A 3 – Pasteurization promotes carelessness and discourages the efforts to produce clean milk.
   - A 4 – Pasteurization would remove the incentive for producers to deliver clean milk.
(3) Food irradiation will preferentially kill "good" bacteria and thus encourage greater growth of "bad" bacteria.
   - B 2 – Pasteurization destroys the healthy lactic acid bacteria in milk, and pasteurized milk goes putrid instead of sour.
   - E 3 – Pasteurization, by eliminating tuberculosis of bovine origin in early life, would lead to an increase in pulmonary tuberculosis in adult life.
   - E 2 – Imperfectly pasteurized milk is worse than raw milk.
(4) Irradiated foods are devitalized and denatured.
   - B 1 – Pasteurization influences the composition of milk.
   - B 4 – Pasteurization destroys beneficent enzymes, antibodies and hormones, and takes the "life" out of milk.
   - D 2 – Pasteurization significantly lowers the nutritive value of milk.
   - D 4 – Infants do not develop well on pasteurized milk.

(5) Irradiated foods are unnecessary in today's food system.
   - C 2 – Pasteurization is not necessary in a country where milk goes directly and promptly from producer to consumer.
   - E 4 – Pasteurization is unnecessary, because raw milk does not give rise to tuberculosis.
(6) Food irradiation impairs the flavor of foods.
   - D 1 – Pasteurization impairs the flavor of milk.
(7) Food irradiation fails to destroy bacterial toxins in foods.
   - E 1 – Pasteurization fails to destroy bacterial toxins in milk.

It is quite clear that the above complaints concerning food irradiation are identical to those voiced about pasteurization almost 100 years ago. As with pasteurization, the charges leveled against food irradiation have no real basis in fact. They are simply a set of unfounded, negative predictions that could be made about any new food process development. Each objection must be considered carefully.

Food Irradiation Will Be Used to Mask Spoiled Food

Spoiled food is spoiled – period. You cannot make spoiled food fresh by irradiating it, just as you can't make soured milk fresh by pasteurizing it. Food irradiation and pasteurization are not employed to mask spoilage, they are employed to prevent spoilage.

Spoilage in this case usually refers to the effects of high levels of microbial contamination. It must first be understood that contamination doesn't simply refer to high levels of bacteria. Yogurt, certain cheeses, fermented raw sausages and certain other foods have incredibly high levels of bacteria. It is only offending bacteria that exert particular negative effects upon the food and thus make it contaminated. These effects normally take three forms:

(1) The contaminated food loses its edibility. The taste, texture, smell or appearance go off. When this happens, nothing can improve the food.
(2) The contaminated food can transmit disease. Pathogenic microorganisms can only transmit disease when they are viable and therefore offensive. When they cannot transmit disease, the bacteria are not offensive and the food is not contaminated. The mere presence of microorganisms in a food at some point in its history does not constitute contamination, or every food would have to be considered

contaminated. It is the presence of the bacteria in an offensive form that can cause negative effects when the product is consumed. Pasteurization is employed to ensure that pathogenic and spoilage bacteria in the milk cannot exert their negative effects. Pasteurization thus prevents spoilage or contamination, it doesn't hide it.

(3) The contaminated food contains toxins. Certain bacteria produce toxins that can result in illness. Once produced, it does not matter whether or not the remaining bacteria can cause infections. The toxins may continue to be toxic. Certain toxins such as the botulism toxin are very easily destroyed by heat. Many toxins, however, are not easily destroyed by heat or other means. If a process removes or destroys the toxin, the food is no longer spoiled or contaminated. If a process does not remove or destroy a toxin, the food remains contaminated.

Canning, freezing, cooking, drying and all other forms of food processing (including food irradiation) have always been considered potential tools to cover up spoilage, when their true function was, in fact, to prevent it. However, there have been occasional attempts to mask or hide spoiled or contaminated food from the very beginning of commercial food processing. The sale of spoiled or adulterated food has been a constant fear and preoccupation with us ever since we left the farm and started to depend upon others for our food supply. Historically, food adulteration was commonplace, and was the catalyst which resulted in our modern food laws. The possibility of adulterated food being sold to the public will always be with us, but that is not the stimulus for the development of new technologies. Frauds will always be a part of our society, which is why legislation is established. In any event, those involved in the food industry know that food irradiation provides no advantage in this form of malpractice over any other food process.

One must differentiate between adulteration which can hide contamination and the processes which remove or prevent offending contamination. Pasteurization does not adulterate milk by killing pathogenic bacteria. If milk has soured and then someone devised a flavor to mask the sourness, that would constitute adulteration.

The best protection against such abuse is effective food laws, backed up with consistent enforcement and heavy penalties. Where such a situation exists, little abuse occurs, regardless of the dire predictions that pasteurization would be used to this end. Even the most miserly of business people realize that fraudulently selling spoiled food can ruin their reputations and put them out of business. Defrauding the public is a bad investment.

Food Irradiation Will Discourage Strict Adherence to Good Manufacturing Practices

Here again, as with pasteurization, there is the presumption that food irradiation will allow manufacturers to dispense with good manufacturing practices. Quite the contrary. When foods are destined for further processing, particular attention must be paid to the high quality of the product throughout the process to ensure that the consequent value-addition will be realized. The more steps in the process, the greater the possibility for product rejection anywhere along the route. An additional step, such as irradiation, requires a stricter adherence to good manufacturing practice so that the products reach the final stages at the highest possible quality level.

Water must be cleaned up and purified before it is chlorinated, because it is simply not feasible to treat anything but the highest quality product. And with reference to the old complaint that pasteurization would end up discouraging good manufacturing practices, it should be known that the prepasteurization quality of milk is higher now than ever before in our history. In fact, the typical souring bacteria are almost a problem of the past, and the present concerns are mainly with cryophilic (cold tolerant) bacteria, which relates to the next concern.

Food Irradiation Will Preferentially Kill "Good" Bacteria and Thus Encourage Greater Growth of "Bad" Bacteria

This complaint, which was also leveled at pasteurization, has some basis in fact. All microorganisms have particular sensitivities and resistances towards physical and chemical conditions. Different varieties of bacteria can have greater resistance to heat (thermophilic), salt (halophilic), cold (cryophilic), acid (acidophilic) or irradiation (radiophilic?) than others. Fortunately, most pathogenic bacteria show very little resistance to irradiation. In certain instances, such as the preservation of vegetables through fermentation, irradiation allows the required lactic acid bacteria to predominate over undesirable species [213]. Bacterial spores are, however, resistant to a wide variety of conditions. Although irradiation increases the heat sensitivity of *Clostridium botulinum*, spores, for example, are resistant to both heat and irradiation [214]. This only points to the need for handling all perishable foods with care, whether they are pasteurized, irradiated, cured or frozen. The concern is generic, and not specific to food irradiation.

This complaint also relates to the one on aflatoxin and has received considerable quasi-scientific attention by New Wave advocates. It has been claimed that studies have shown that irradiation results in higher levels of aflatoxin production in foods [115]. Two studies carried out in 1976 and 1979 were quoted. In these studies, the foods were first irradiated, then sterilized by heat, then inoculated with mold spores and then incubated [116,117]. These are not conditions that are realistically encountered anywhere in practice. Conveniently neglected was the great majority of more rationally designed work which has indicated the very opposite effect [118–122,215].

Another example of this issue elevated to a masterpiece of perverted logic appears in the book *Food Irradiation: The Facts* [123] and is repeated with minor variations in *Food Irradiation: Who Wants It?* [124].

> Currently some importers go to considerable lengths to ensure that imported nuts are not contaminated with aflatoxin-containing bacteria. They fear that if irradiation is legalized, suppliers may irradiate nuts to keep the bacterial count below the control levels. In doing so, the competitor bacteria will also be destroyed and within a short period after the importers have purchased what they believe to be "clean" peanuts, aflatoxin production could rapidly increase. It is the processors in this country not the foreign suppliers who will be blamed by the consumer for any subsequent health hazard.

In the first instance, the reader should know by now that aflatoxin is produced by a mold, *Aspergillus flavus*, and not a bacteria. Nuts are regularly inspected for aflatoxin (and other mycotoxins) as well as for molds. Mold-containing nuts are often separated from the rest by modern color sorters. It is most often only a few nuts out of a batch that are moldy. If those are removed there is little chance of aflatoxin (which is why you should not eat any discolored or moldy-looking nuts). When a batch is made into peanut butter or meal, the aflatoxin from any infected nuts that may still be present is spread throughout the batch.

Any tampering with the nuts which may result in higher aflatoxin levels would be detected in the mycotoxin analysis and be self-defeating for all concerned. This whole concocted fantasy is simply another absurd attempt to show that the goal of all, except the New Wavers, is to victimize the consumer.

Irradiated Foods Are Devitalized and Denatured

The concept of a life- or vital-force in foods is as old as man. It is a primitive idea that has been experienced by all tribes, stone-age and modern [125]. Foods are not alive in the strict sense. They may still have enzymatic activity which allows certain metabolic functions to take place, but all natural, unprocessed foods begin to perish the moment they are harvested. A typical example is the dramatic loss of vitamin C (ascorbic acid) from salad leaves. It has been reported that up to 18% of vitamin C is lost within two hours of harvesting the green leaves and reaches as high as a 66% loss after ten hours – a typical delay before they are purchased [126] at the market. Aside from the natural loss of nutrients that results from physiological spoilage, freshly harvested foods become subject to attack from a wide range of spoilage microorganisms.

With the exception of intentionally fermented foods, such as yogurt or soy sauce, or sprouted seeds, *life* is not a quality that is necessarily desired in foods. In fact, the main purpose of food preservation is to halt or slow down the live spoilage process (microbiological or enzymatic) in foods, while retaining the maximum nutrients and eating qualities. This idea of preservation has been with us for a long time. It has proven to be more useful than the idea of a life force, which is why people no longer eat an animal's heart or brain to become brave or smart.

There has been much said about nutrient losses along with the false notion that irradiation is the only process where this occurs. This is not the case. Although it is occasionally possible to increase the availability of certain nutrients through irradiation [216–218], all forms of processing usually lead to some nutrient losses. The most basic staple of our diet – bread – is a processed product of wheat. The nutrient losses which occur during the conversion of wheat into bread – even whole grain bread, are very significant. Milling of wheat reduces the physiological effects of the fiber content. Vitamins are lost during baking. The most critical essential amino acid, lysine, suffers significant loss through the

Maillard (browning) reaction which results in the golden crust. However, if it were not for baking or other similar heat treatment, wheat could not be consumed, and would not be available as a food, and Western civilization would not have evolved as it has. Raw wheat is inedible. You have to process it and destroy some nutrients just to eat it.

The same goes for milk, unless one wants to risk certain diseases. The same goes for meat. The same goes for canned or frozen fruits and vegetables if the goal is to preserve them for a long period after harvest. The question is whether to make do with some nutrient loss or not to make do at all. If on the other hand, you wish to derive the maximum amount of nutrients from meat or fish, you can always eat them raw and risk the possible consequences. Many do. If that doesn't suit, one must be prepared to make rational compromises.

Irradiated Foods Are Unnecessary in Today's Food System

Today's food system, like yesterday's food system, requires improvement. It is not necessary for all consumers to purchase poultry, beef, pork, fish or dairy products that contain pathogenic bacteria. It is not necessary for all consumers to contract foodborne diseases. It is not necessary to use chemical treatment for all types of insect disinfestation. It is also not necessary to pick strawberries and other fruit when they are green and tasteless. For those people who prefer such foods, they should be available. For other consumers who are not satisfied with these foods, safe and proven alternates are a necessity, and must be provided.

Food Irradiation Impairs the Flavor of Foods

This complaint stems from early research work in the field of food irradiation. Researchers were testing every type of processing condition and dutifully reporting all the results, good and bad. This is the normal course of scientific progress. Some products tasted poorly after processing. They were reported in the literature. With changes in the times, temperatures and other conditions of treatment, this negative impact was eliminated or minimized.

Referring to such experiments as a final proof of the negative effects on taste is no different than a neighbor observing your first attempt at making a soufflé. It probably collapses on the first attempt. You keep on trying. Eventually, you work out all the conditions so that you become the very best soufflé maker in town, but this neighbor will always remember, and refer to your one soufflé that fell.

It should also be borne in mind that most of this initial research was carried out by research institutes, and not by the food industry itself. It can be taken as a given that the food industry will not and cannot accept the production of foods with poor flavor [219]. Industry must be competitive and viable. Making foods that taste bad doesn't achieve either goal.

This is a silly and somewhat desperate complaint. If consumers do not like the taste or flavor of foods, they won't buy them. Any food manufacturer who does not know this, is in the wrong business. Any New Wave food advocate who doesn't know this is also in the wrong business.

Food Irradiation Fails to Destroy Bacterial Toxins in Foods

Again, this is also a somewhat desperate complaint. Neither heating nor irradiation will destroy certain bacterial toxins, such as the staphylococcal enterotoxin. This is known at the most elementary levels of food technology and the food industry. It has long been known because we have been heat processing foods for a long time. Any food contaminated with bacterial toxins cannot be processed to hide the toxin. Any complaint in this regard which is leveled at irradiation can be equally leveled at cooking, pasteurization or canning. Bacterial toxins result from high loads of toxin-producing bacteria. The food industry (which has much at stake), knows well that the best approach to bacterial toxins is the prevention of their formation through hygienic manufacturing practices and high product standards. Food irradiation has been found to be effective in reducing toxin loads through this method of action [220]. The food industry does not make money from sick or dead consumers — they want their clients alive and well and hungry.

The complaints do not appear to be specific to irradiation, because they have also been used as a basis of complaint against pasteurization. If these original complaints had continued to influence politicians, legislators, manufacturers and retailers, pasteurized milk would remain unavailable as a food to consumers to this day. It would not be part of a free and informed consumer choice. Fortunately, but not without considerable useless and harmful delay, scientific fact prevailed over fiction, and pasteurized milk has become a central part of our food choices. This situation has not yet come to pass with irradiated food. The myth-information persists and the deliberate association of food irradiation with incidents such as Chernobyl make it all too easy for the myths to be perpetuated.

Aside from the above generic complaints, which could be directed at any food preservation technology, there are other complaints which are specifically aimed at irradiated foods.

Irradiated Food May Be Radioactive

This is an understandable fear to people who have no knowledge of food irradiation and how it works. Radioactive material is not deposited on the food. Any radioactivity which exists in the food is the natural background level which was there before any processing ever took place. This natural level exists in all foods and in all things.

Theoretically, it is possible to induce radioactivity in foods, but from a practical point of view it is virtually impossible. At the energy levels available from the sources employed for food radiation, no radioactivity above the natural background levels has ever been detected even when the products are over-processed. The most definitive statement to this effect was issued by the U.S. Food and Drug Administration in the *Federal Register* of 1986:

> Because no evidence has been submitted to contradict FDA's finding that the irradiation of food does not cause the food to become radioactive, no further discussion of this issue is necessary. [127]

However, this has not prevented the perpetuation of the myth that irradiation of food may make it radioactive. In both the British and American editions of their book on food irradiation directed at consumers, Webb and Lang (and Tucker in the U.S. edition) [123,124] admit that it is unlikely that the irradiation process will make food radioactive, but nevertheless, they describe the measurement of treatment levels in the following manner:

> In measuring the dose to food we are concerned about the amount of energy that has been *deposited* as a result of irradiation. (Author's italics)

Although the use of the word *deposited* was most likely not intentional, it does leave some readers with the impression that something extra was left in the food. It is an odd description for the absorption of energy. It would be very unusual to describe the heat used to boil an egg as being deposited on it. Thus the particular language used, even if it was unintentional, results in the myth not being fully discounted. In any issue which is not well understood, it is imperative to use language which does not leave false impressions.

In 1989, the Australian government placed a three-year moratorium on food irradiation. As the expiration of this moratorium was imminent, Australia's National Food Authority drafted a single proposal, in two parts, for the regulation of irradiated foods. The first part of the proposal dealt with the actual use of food irradiation, i.e., whether it should be permitted and, if so, under what conditions. The second part of the proposal indicated the National Food Authority's intention to develop a standard for foods accidentally contaminated with radioactive substances (e.g., cheese from Chernobyl). Although one issue has nothing to do with the other, they had both been intentionally linked together in the same proposal. The net effect of linking the two issues in a single proposal is to perpetuate the myth that irradiated foods may become radioactive. The proposal thus misleads all those who do not have a technical understanding of the issues. If the National Food Authority of Australia can issue a proposal linking food irradiation with radioactivity in foods, then there is little wonder that the myth persists, despite the above-cited Food and Drug Administration statement.

Irradiation Brings about Harmful Chemical Changes in Foods

This complaint refers to the formation of free radicals and radiolytic products as previously discussed. Most natural processes and transformations go through a free radical stage and form what are referred to as *lytic* products. I suppose you could refer to such products formed as a result of cooking as *thermolytic* products, or freezing as *cryolytic* products. It is even possible to produce this effect with high frequency sound, in which case the end results would be referred to as sonolytic products [221,222]. The important issue is not the source of their formation, but whether or not they are harmful. There is no evidence that the radiolytic products in irradiated foods are harmful, despite a far greater degree of testing than any other food process. Irradiated foods have proven to be safe. On the other hand, the pyrolytic products formed at much higher levels during the broiling, smoking or frying of meat or fish or the roasting of coffee can be mutagenic or carcinogenic [128 – 130]. If pyrolytic products from broiling may be harmful, and radiolytic products from irradiated foods are not, what, then, is the issue? Why are radiolytic products rather than pyrolytic products focused upon by these advocates if they are supposed to be concerned for the health of consumers?

Irradiated Foods Can Cause Genetic Mutations

Since there is a confusion of irradiated foods with radioactive materials, there is a built-in fear that these foods could cause genetic mutations. Once the facts are known that irradiated foods are not radio-active, why should they be considered more capable of causing genetic mutations than any other foods? There is no reason for this. But, as we are all human, we don't necessarily think of everything in an analytical way. In addition, if one is not too familiar with the subject, it is all too easy to imagine that irradiated foods can cause genetic mutations.

Polyploidy

The anti-irradiation lobbyists were handed a bonanza in the mid-70s when scientists from the National Institute of Nutrition in India reported genetic abnormalities in malnourished children fed freshly irradiated wheat [131]. The same institute also carried out work on mice, rats and monkeys with similar results [132−134].

The reports stated that consumption of freshly irradiated wheat resulted in a condition called *polyploidy*. When human lymphocytes (white blood cells) are viewed under the microscope, forty-six chromosomes are normally present. Polyploid cells have twice and sometimes three or more times the number of chromosomes. To complicate matters further, the medical significance of polyploidy is not really known. Healthy humans normally have an incidence of about 0.5−2% polyploid cells in their lymphocytes. This means that when you look at 100 normal human blood cells under the microscope, about one cell on average will show up as being polyploid. The words on average are emphasized because if you count only 100 cells, your particular sample could be missing one polyploid cell or you might possibly find five or six such cells. You only need to be off by one cell to show a full percentage point difference. Since there is only about 1% polyploid cells in human lymphocytes, you really should count many more than 100 cells if you are serious about getting a true average. To get an accurate idea of the true average number of polyploid cells, a large sample of cells (in the thousands) should ideally be counted. (This does not require a larger blood sample, as they will all be in the same single drop of blood. It just takes longer to count them.) Based on such studies, it can be stated with confidence that humans have a normal incidence of approximately 0.5−1% polyploid cells.

When analyzed, the work at the National Institute of Nutrition revealed a number of peculiarities. The malnourished children who were fed freshly irradiated wheat for one month were found to have an incidence of polyploid cells of about 0.8%. The malnourished children continued to be fed freshly irradiated wheat for one and a half months and eventually had a polyploid cell incidence of 1.8%. The malnourished children who were fed unirradiated wheat throughout the study showed a 0% incidence of polyploid cells.

When the report is examined more closely, it is found that only 100 cells from each of the five children in each group were counted−an incredibly small sample upon which to base any conclusion, particularly when the blood samples already taken could have yielded much more reliable results if they had been studied more thoroughly. Furthermore, although the results in each group were averaged, there is no indication of the actual incidence in each child. In order to arrive at 0.8% for the one-month test, did four out of five children have one polyploid cell each from the total of 500 cells counted? Or did one out of five have four polyploid cells? The report does not reveal this. In the next group at one and a half months, nine polyploid cells were found in the 500 counted. Did one child have nine or did four have two each with one in the last sample? Again, no clarification. One might say that this issue is not so important, the important thing is that the levels reported from these samples were greater than from that of the children fed unirradiated wheat. This idea, however, led to another problem.

Curiously, the levels reported from blood samples of children fed unirradiated wheat were 0%. An incidence of 0% is almost never found under normal circumstances [135]. In fact, you would expect to find much higher than normal levels in children who have suffered severe malnutrition. The chromosomes of malnourished children are hard to analyze unless one is truly expert. Malnourishment results in a large number of fuzzy-looking chromosomes which are difficult to characterize [136]. None of this was taken into account in the published reports.

The design of the study, the small samples used, the lumping together of results and the abnormally low figure for their control group do not provide a basis for a rational conclusion. The published report revealed nothing about the actual diet given that would allow other scientists to reproduce, and possibly confirm, the study. It says the children were fed irradiated wheat together with skim milk powder, sugar and clarified butter. Was the wheat ground into flour? If so, was it whole wheat flour, white flour, bulgur, or was the whole diet mixed into a dough and made into a characteristic Indian fried bread? If so, was it fried once and kept for four weeks, or made fresh daily, or something in between? The published report says absolutely nothing. The report states that all the "freshly" irradiated wheat was fed within three weeks of irradiation, but the test went on for six weeks! This means that at least two batches or more must have been prepared. What were the children fed during week three? Was it one-, two-, or three-week old wheat? A legitimate scientific report must give this information, and the procedure must be consistent, if it is to be subject to verification. This report was clearly not an exercise in controlled observation, and it was thus not science.

Taking young starving children and putting them through a test without knowing the possible outcome raises some serious moral issues. (There is no indication in the report of consent of any kind given for these tests.) Obviously, one should never take possible risks with starving children. Once it was done, then at the very minimum, out of respect for the children, it should have been done correctly. This was not the case. The work was poorly done, and very poorly reported. The subjects were not a group of laboratory rats that could be retested over and over. These were children, who should not have been chosen in the first place. Even if rats were used, the science should have been better. An internal review of this work was requested by the Ministry of Health of the Indian government shortly after it was published. The review report repeatedly states that the results suffered from bias and preconceived notions on the part of the scientists. Interviews with the scientists revealed that one of the authors of the report (who was the senior researcher) was very doubtful of the results because the fuzzy chromosomes made cell counting so difficult. He even went so far as to say that the data were so imprecise that no importance should be attached to them.

The results of the original publication were studied by scientists all over the world and the conclusion reached was that the study was simply unacceptable for the purpose of drawing any conclusions at all. The issue, however, inspired other scientists to see if any relationship did exist between polyploidy and the consumption of irradiated foods. Tests were carried out in China with consenting adults, and on much larger cell samples, but the results clearly indicated that consumption of irradiated food does not cause polyploidy or any other genetic aberrations whatsoever.

A recent defense of the original report [137], by authors in the same institute, states that "it is difficult to escape from the feeling that all findings which are in favor of the wholesomeness of irradiated foods are readily accepted, while observations which raise doubts and question this stand are either viewed with suspicion, either covertly or overtly, or outright rejected." This is an extraordinary statement to appear in a scientific journal. It implies that there is a worldwide conspiracy among scientists to promote irradiation. Yet the defense doesn't answer the scientific questions about the poor quality of the original work and its reporting. It still doesn't tell anyone if the wheat fed the children was in the form of raw grains or sweet chapatis. It doesn't conclusively comment on the medical significance of polyploidy. The fact that the children fed the freshly irradiated wheat had better weight gain, better serum albumin contents and better hemoglobin levels indicated that they were recovering faster during the test. How would this affect interpretation of results? The publication doesn't say. It simply concludes with the implication of a conspiracy.

Despite the repudiation of the original work by the international scientific community, and even one of the original authors, anti-irradiation New Wavers continue to make constant reference to it. They do not refer to the unacceptable shortcomings of the study, nor to consequent work indicating that there is no relationship between the consumption of irradiated food and polyploidy, because this would not serve their purpose. In one of their frequent appeals for money, the New York-based Food and Water group stated that 80% of the children in this study appeared to be developing cancer. Using shoddy, unproven notions on the one hand, and the universal fear of cancer on the other, has proven to be a very effective means of prying money out of the public.

Even the very logic of feeding freshly irradiated wheat is flawed. In order for the process to be useful, wheat or any other product for that matter, should be irradiated as soon as possible after harvest in order to minimize losses. The harvest must be stored for the whole season. The wheat should be irradiated and then stored until needed by the mills. The older the wheat, the better the quality, because wheat needs to be mature in order to perform best. (The sulfhydryl or $-SH$ groups in wheat protein must be oxidized to disulfide groups $-S-S-$ in order to give the maximum elasticity to the doughs.) After storage, which could be for several months, the wheat is milled and then stored again, although for a shorter period.

The New Wavers do not explain to their followers that since the entire basis upon which the polyploidy premise was originally made has proven to be totally unreliable, then there is no burden of proof remaining. If the original premise is false, no one should have to disprove it. No tests must be made to disprove something that was never proven in the first place. On the contrary, the burden of proof must be placed upon those making the inference that there is a proven relationship between the consumption of freshly irradiated food and the incidence of polyploidy. If they wish to properly establish their premise, they must carry out a repeat experiment with the depth and precision required of scientific observation. Until this is done, the polyploidy issue is nothing more than a phantom. Its only significant result was the creation of a full-blown myth and the promulgation of a watershed of myth-information.

The Food Irradiation Issue Needs a Properly Referenced Scientific Report

Some New Wave advocates who are against the use of food irradiation claim that there is a need for a fully referenced report on the use and safety of the process. It is understandable that if one wanted to independently judge any issue, such a report, complete with references, would be useful. They state that no such report exists and, by implication, foster the notion that no proper scientific judgment is possible at this time. They do not accept the judgment of the acknowledged scientific authorities, and make the presumption that they have not done their homework. The New Wavers also make the assumption that they themselves are qualified to make such a judgment, even though they do not have the professional training to do so. There can be no greater proof of this than the muddle they have made of the information available.

To be placed into perspective, it must be understood that the New Wavers who claim the lack of such a report are not experts in the fields of food processing, irradiation or health, nor are they practicing scientists in a related field. If they were, they would realize that scientific literature is replete with more studies and reports attesting to the value and safety of irradiation than any other single process. One recent textbook by a recognized world authority on the subject is entirely devoted to the singular issue of the safety of irradiated foods [9]. There are no such similar reports attesting to the safety of canned, broiled, smoked, microwaved or dried foods. All the information on food irradiation has been published and reviewed by professional scientists. It is also readily available in journals and centers of research. Consolidated reports of available literature are costly and take considerable time to prepare. Those that specifically demand such a report must be prepared to put in the effort themselves, or else be willing to fund its preparation.

The Lack of Difference between Irradiated and Nonirradiated Foods Will Encourage Abuses

The fact that food irradiation has so little effect on the processed food is certainly not a reason to fault it as a processing technique. In fact, it is the ultimate goal of the food technologists to have processes that could improve a food without affecting its consumer acceptability in any way. The ideal foods would, of course, be freshly harvested and natural, throughout the year. This is not possible. If it were a practical and competitive possibility, we would all be eating freshly harvested, natural foods, on a regular basis. The fact that we don't is confirmation that there are foods which, for reasons of availability, acceptability, appearance, cost and safety, we prefer processed in some way. The fresher or more natural the foods appear, the more desirable they generally are to the consumer. We do the same thing at home by packaging foods or storing them so they will be edible and look acceptable for the longest period of time.

It is curious that the advocates who say consumers will not be able to identify irradiated foods are the very same people who say consumers will not accept irradiated foods because they taste, feel and look "off."

In the food industry, an incredible amount of work goes into the development of technologies that make foods more convenient and appropriate for our changing requirements. In the development of these technologies, every effort is made to minimize the effects of processing on the food itself. Irradiation improves the quality of foods with no obvious effects. In fact, this is one of the great advantages of food irradiation. It can accomplish an important function without changing the basic character or acceptability of the food. It was this very trait that made pasteurization so valuable. Yet, this advantage is held against food irradiation by the New Wavers, who claim it will promote rampant abuse throughout the industry.

If there is concern that foods may be irradiated without consumers' knowledge, then there are certainly ways to handle this within the current food management systems. Legislation and full-time inspection, as in the meat industry for example, will work equally well with food irradiation. As in any other industry, there may be an occasional abuse, but heavy penalties and poor publicity have always served to control violations.

It is somewhat ironic that those who purport to represent the rights of others would put forth the presumption of abuse. It is a basic premise in democratic societies that the ability to commit an abuse or penalty does not lead to the presumption of guilt. If not, we would all be accused of crimes before we ever thought of committing them.

To fashion the ability of food irradiation to process foods that have a high acceptability rating with consumers into a complaint against the technology is both immature and unjust. It is tantamount to saying that we should go back to boiled, rather than pasteurized, milk because consumers will be sure to identify it by its burnt taste and thus be fully certain that it has been processed.

There Is No Satisfactory Test for Irradiated Foods

Here again, the presumption is made that there is something wrong with irradiated foods, and there will be a need to have a test to verify if a food has undergone treatment. The same complaint was made against pasteurization. In fact, it would be very useful to have a test to determine if a product has been previously irradiated, and recent research indicates that such analyses based on electron spin resonance, luminescence and a variety of other methods will be available shortly [138–140,223–230]. But this is no reason to delay the introduction of irradiation. There are many areas where tests could be useful to determine if products have received certain previous treatments, but the lack of such a test has never prevented the employment of the technology. Freezing is a perfect example of this.

According to Webb and Lang [123], the position on irradiation taken by the British Frozen Food Federation after publication of the Advisory Committee on Irradiated and Novel Foods (ACINF) Report is (with reference to the issue of a test for irradiated foods):

We recommend that:

a) Food irradiation should not be legalized until a simple test, which indicates whether irradiation has been applied, is generally available. We strongly recommend that Government affords funding to a suitably qualified research establishment to ensure such a test can become available before legislation and that its simplicity is at "litmus" paper level.

This is a rather opportunistic and hypocritical statement. Irradiation may be a competing technology to freezing in some cases, and such statements must therefore be taken in that light. But what about freezing itself? Consumers should know if the "fresh" fish or meat they see in the supermarket has been previously frozen, should they not? What simple, litmus-paper-level test is there to determine if a product has been previously frozen? There is none. Some products, such as strawberries, don't freeze well, and consumers easily know when they have been frozen, simply by their poor appearance and texture after defrosting, compared to the fresh product. But many products freeze very well. How can consumers tell if the product was frozen? There is no simple test to alert those consumers who don't want frozen foods.

What about products that have been frozen, defrosted accidentally and then frozen again? Repeated freeze-thaw cycles are an important problem in the supermarket industry. How are consumers to know if products have been abused by accidental defrosting and refreezing? Again, there is no test for the product itself. The technology for cheap indicators, which show if a frozen product has been defrosted somewhere along the line, has been around for almost 20 years, but neither the freezing nor supermarket industry has made it available to consumers. Freeze-thaw indicators are small enough to be applied to the individual retail consumer packages, but consumers still don't have access to them. Thus, a product can be subject to repeated freeze-defrost cycles, either at the freezing plant or somewhere in the distribution system, or in the supermarket, and there is still no way for the consumer to be aware of this. Manufacturers don't even include a small, odd-shaped or colored ice cube which would melt and allow consumers the chance to see if a product was defrosted before it got to them. Why not? If one is to look for potential abuses, then a good look should be taken.

There is also no test to determine if previously frozen foods have been used in processed products. When you buy frozen pizza, you have no way of knowing if the mushrooms or seafoods used to make it were previously frozen. Your only knowledge of previously frozen products being served at a restaurant is through the good will of the restaurateur.

Should freezing be made illegal until a simple test is established? Of course not. There are countless examples of physical processes being employed without a specific test to show that they have been applied.

The concern for a test really reflects a worry on the part of some consumers that if irradiated foods are not labeled properly, there will be no way to know if they are irradiated. In fact, this problem exists in other food processing areas. Where no test is available, thorough government audit procedures, such as full-time inspectors working at plants, are a normal measure until analytical tests are developed.

Irradiation in Combination with Other Processes Will Reduce the Nutritive Value of Foods

There is a concern that irradiated foods may be further processed, and that this will lead to even greater nutritional losses in foods. While it is true that combined processing can diminish certain nutrients, this concern is equally applicable to all types of processing. At times, the opposite effect is noticed – that is, an improvement in nutritional value of the food if the combined process releases previously bound nutrients. An example of this would be the greater availability of calcium in frozen and then canned fish as opposed to fresh fish, because the combined process releases calcium from the bones. Irradiation combined with heating, such as baking with irradiated flour or cooking irradiated legumes, has resulted in the greater availability of certain key vitamins such as niacin, thiamin and riboflavin [141]. The most recent reports have indicated that combination treatments are not a practical concern for food irradiation any more than for other processes [142,231,232].

Irradiated Foods Will Not Be Properly Labeled

The issue of labeling irradiated foods is a rather interesting one. At the outset, the first thing to understand is that irradiation is a physical process – nothing is added. Furthermore, food irradiation leaves no residues in the food. Typically, any such process would not be required to be specifically labeled for consumers. This is unfortunate, because many consumers would like to know how their products are processed from beginning to end. For instance, many products such as prepared tomato sauces are made with tomatoes that were previously canned, yet there is no such indication on the label. Soups or prepared meals are often made with frozen meats or fish, but again, there is no indication of this anywhere on the label. Raisin bread, fig cookies, or other fruit-filled baked goods are made with dried, frozen, cooked and frozen fruit, yet the consumer is not informed of any preprocessing prior to the production of the finished product. Some instant pastas, such as those in instant soups, can be made by microwave drying, but no such indication will appear on the label. Even when fruits, vegetables or spices are treated chemically (such as bananas ripened with ethylene gas, spices decontaminated with ethylene oxide, or potatoes and onions treated with isopropyl-n-chlorophenyl carbamate or 1,2,4,5-tetrachloro-3-nitrobenzene to prevent sprouting), no labeling disclosing pretreatment is required for consumers—though residues may still be present. Consumers who pay for these products have a full right to know how they have been prepared. For example, certain soft cheeses are made with raw, unpasteurized milk. If consumers were aware of this they might choose not to consume them.

Fortunately, there is a general agreement by all concerned that irradiated foods should be labeled as such. The U.S. government has officially stated that irradiated foods should be labeled, "not based on any concern about the safety of the uses of irradiation," but because of the significance placed on labeling by consumers [143].

This understanding is not based on any of the traditional reasons for the need to label food, but rather on the need to ensure that a process which has received so much controversial attention is readily apparent to consumers. In a sense, however, irradiation as a process is discriminated against. Foods that are not specifically highlighted through labeling do not alert consumers. They do not elicit a conscious decision regarding purchase. This discrimination is not considered a negative, however, since irradiated foods can offer a greater degree of safety from foodborne diseases, and it is felt that consumers may, in fact, seek out these foods. It is also felt that promulgation of the debate over the labeling issue would only serve to lessen consumer confidence. Thus, from a labeling point of view, irradiated foods will respect the needs of consumers to be fully informed about what they purchase. This is more than many other foods currently on the market do.

The proposed form that labeling will take is another issue that is often brought up in the food irradiation controversy. Those who favor irradiation want an easily recognizable symbol, such as the Radura symbol (Figure 6.1), together with a written notation that the food has been

FIGURE 6.1 The Radura symbol.

irradiated. The goal is that once the symbol is fully understood by consumers, they will immediately recognize irradiated products. The hope is that, with time, it will be recognized as a symbol of quality much like the Woolmark, or Good Housekeeping symbol. The intent is that the consumers should know that the food has been irradiated.

Opponents of irradiation insist that only a written indication, or preferably a warning, that the food has been irradiated is acceptable. They feel that a symbol, such as the Radura symbol, will mislead the consumer. They imply that such a symbol will be used by the industry to hide the fact that the food has been irradiated, and thus deceive the consumer.

Symbols are the most effective means of communication. A bottle with a skull and crossbones on it is immediately recognized as poison. Symbols surround us. It is because of symbols that we can quickly find our way to the airport in any country, regardless of language, simply by following the airplane highway signs. Because of symbols, men do not accidentally wander into ladies' bathrooms. I have even seen symbols of men's and ladies' hats replace the typical gender figures. Symbols provide instant recognition. Symbols are not the issue. The issue is that the symbols must be fully understood by consumers.

If the intent is to have consumers recognize irradiated food products instantly, and understand that there is a lower risk of foodborne diseases, then a symbol is definitely the way to do it. It must be a positive symbol with accompanying text to ensure that it is understood. The use of the symbol should at least be optional, with a written statement compulsory.

Food Irradiation Is Dangerous to Workers and Will Cause Environmental Damage

This complaint does not concern the safety of irradiated food per se, but is nevertheless an important issue. Within the last three decades, there has been an increasing international concern over the deteriorating condition of our global environment. If we are to leave our descendants with an environment that is sustainable, it will be necessary to dramatically change some of our ideas on how to manage our resources, and how we treat our planet and its inhabitants.

Food irradiation has been brought into this important issue primarily because of its "radiation" connection. While it is beyond the scope of this book to delve into the whole issue of environmental sustainability in any comprehensive manner, the relationship to irradiation can be examined. In doing so, the differentiation between food irradiation and nuclear energy or nuclear weapons must always be borne in mind. Food irradiation cannot be confused with either, if any meaningful understanding of the issue is to be had.

The key aspects to be considered are the transportation of ionization sources (Co^{60}), the general safety of facilities, and what can possibly go wrong that might affect the environment. Fortunately, there is a long and established use of irradiation in the sterilization of medical devices such as syringes, bandages and pharmaceuticals, and the treatment of other materials such as plastics, packaging and wine corks, upon which to make these assessments. (Even the blood used for transfusions is irradiated.) To a lesser extent, the same can be said about the Co^{60} treatment units that are standard equipment today in many hospitals around the world.

Most of the environmental concerns apply only to food irradiation facilities which use Co^{60} as a source of ionizing radiation. High-energy electron beam food irradiators can be turned on and off with a switch, and thus pose no more potential environmental damage than a baker's oven does.

Co^{60} sources are made under very closely controlled conditions. Once ready for shipment, they are placed in lead-lined containers that are designed to withstand incredible abuse. These are officially called IAEA Type B (U) containers. Among the situations which they must survive are thermal conditions equivalent to thirty minutes in an environment of 1472°F (800°C). Other tests carried out in the U.S. and U.K. were a free-fall drop of a container from an aircraft at 2,000 feet onto a concrete runway (which is a very unyielding surface), and the impact of a locomotive traveling at 80 mph (the locomotive was destroyed while the container remained undamaged). Since the beginning of shipments of radioactive materials in North America (starting around 1955) there have been in excess of a million deliveries. Despite the very few incidents of delivery accidents with Type B containers, there has been no radiation exposure to the public or the environment [144]. This record is testimony to the design and production of equipment specifically constructed to carry out a particular function. This is not to say that the accidental release of radiation is impossible. But we do not normally run our lives in the realm of extremely minor possibilities rather than realities. The normal course of action is to engineer for the possibilities and then examine the record. If improvements can be made, they should be.

Irradiation facilities are likewise designed to fully contain the γ-rays emitted by the source. The buildings are designed to withstand earthquakes, and the walls of the irradiation chamber are made of concrete thick enough to contain all radiation. Normally, the ionization source sits in a pool of water when not in use. Its only effect is to warm the water slightly, as a heating element might. Even though Co^{60} is not soluble in water, there are filtration systems in place to ensure that the pool water remains free of any possible radioactive material. The water is periodically tested for potential contamination.

The ionization source is lifted out of the water in order to irradiate the various products. There have been rare cases where the source has temporarily stuck in the raised position. The routine course of action is either to call in a mechanical expert or to fully drop the source into the pool. Aside from possibly ruining a batch of food through overtreatment, there are no other consequences.

Thus, in all the experience gained with irradiation of foods, plastics and medical products, there have been no major accidents resulting in environmental effects. Very few other industries can make this claim.

Since the half-life of Co^{60} is a little over 5 years, it is generally the policy of most commercial irradiators to continually replenish the source pencils in order to maintain a fairly constant level of γ-ray emissions. This practice reduces the need to compensate for the reduced emissions by leaving products to rest longer in the irradiator. Source pencils are shipped back to the original supplier, who generally stores them in a government-approved manner and location.

The issue of worker safety is also extremely important. It applies to all industries. As with foodborne diseases, it is an issue that should only be left to experts and workers. You don't see or feel γ-rays. Workers are therefore equipped with dosimeters or other radiation detection devices to ensure that they are not exposed. Safety standards have been set for worker exposure, and the facilities have been designed to meet these standards. As with all other processes, if evidence is ever produced that such standards do not adequately protect workers, then the design of facilities or procedures will have to be altered to meet newer and better standards. It is not in anyone's interest to have workers at risk. Lawsuits and bad publicity are not experiences which any industry welcomes. Having a well-trained and healthy staff is critical to smooth, consistent, long-term operation. In all the years of irradiation facility operation, there have been very few worker accidents. These were always the result of bypassing safety procedures. Despite the excellent safety record for the operation of food irradiation facilities compared to other industrial operations, any supplementary procedures which could prevent the easy overriding of safety protocols should be promoted.

When all the evidence is taken into account, it is difficult to get a clear view of the intent of the anti-irradiation advocates in this matter. If consumer health and safety were their goal, then there is certainly more than enough evidence to support the use of food irradiation. If consumer health and safety were the priority, then the legitimate concerns of public health officials would not be so quickly brushed aside. In a world where absolute safety is unattainable, there is no choice but to make decisions regarding the risk and benefits of any process. The consumer cannot be protected by hearsay, or moral and political opinions. The consumer must be protected by reliable evidence and experience.

In order to put the whole issue into some type of realistic context, the following should be understood and remembered:

- Foods spoil with time. They can contain disease-causing bacteria, parasites and pests. They can make people sick and can even cause major epidemics. It is a fact of life. This is not meant to frighten consumers, but rather to emphasize that, despite these potential problems, our food supply is surprisingly safe. This is no accident. An army of scientists, technicians, legislators and inspectors are there to ensure this.

- There were no good old days of food. In the good old days, the rate of sickness and mortality from foods was beyond our worst nightmares. In order to manage this problem, objective scientific knowledge and study are employed. It is not sentiment, social beliefs or political principles that prevent foodborne disease. It is the technical knowledge of these diseases and their control that can prevent them. Offensive organisms are neither moral nor philosophical. They are biological.

- Foods that are irradiated or ionized do not become radioactive. They therefore do not expose people to any form of radioactivity.

- Food irradiation produces the same free radicals and radiolytic/ionization products that other forms of food processing do (cooking, canning, freezing, etc.). There may be some unique microlevel-*lytic* products that result from irradiation, just as there are unique microlevel-lytic products that result from all the other forms of processing. There is no basis whatsoever to presume that those formed during irradiation are any different from those formed during all other processes. There has been no significance associated with their presence in foods.

- Food irradiation works by ionizing complex macromolecules, and can thus prevent their negative spoilage effects. The advantage of food irradiation over other methods of processing is its ability to evenly penetrate foods and thus achieve ionization of macromolecules without substantially altering the nature of the food. It thus allows the consumer to prepare the food in whatever manner desired.

- Although fruits and vegetables continue to undergo certain metabolic processes after harvesting, they are not "live" nor do they contain any magical powers. The latter idea went out of style when we decided to drop the practice of cannibalism. Formerly live food deteriorates continuously from the time it is harvested. Preservation techniques are employed, where possible, to prevent this deterioration. The very few intentional live foods, such as unpasteurized yogurt, should no more be irradiated than cooked.

Such treatments will destroy the enzymatic or bacterial activity that is considered desirable. The live nature of these foods dictates rigid control measures to prevent contamination or spoilage by undesirable live microorganisms.

- The nutrient losses experienced during irradiation/ionization are no different than those which occur as a result of all the other common types of processing, such as canning, cooking, drying and bottling. Any possible negative effects from such losses will more than likely be made up by the greater availability of safer foods.

- Food irradiation/ionization is no different from any other food process. Although it can carry out certain functions better than other processes, some other functions are better done by alternate means. Ionization, like any other process, is not the answer to everything, nor was it ever claimed to be. It is simply one of a battery of technologies developed to provide a safe and wholesome food supply, and to keep pace with the necessities of feeding a growing population. Despite the cynicism or complaints expressed about our modern food supply, more people are living longer, more active and productive lives today than at any other time in our history.

CHAPTER 7

Irradiated Foods and the Consumer

Consumer Reaction to Irradiated Foods

Regardless of all advertising, posturing or hypothetical predictions, the ultimate test for any product or process is the marketplace. It is the consumer who determines whether or not a product is better than previous or competitive products. The whole process of making this determination is rather complex and involves social or cultural biases, initial perceptions of benefit or risk, anticipated consequences of eating a food, comparison with alternate products, etc. The collective results of these individual judgments, many of which are carried out unconsciously, are incorporated into the final decision the consumer makes about whether or not to purchase the product.

Not surprisingly, a major effort is made to influence consumer choices at each of the various stages of decision making. Perceptions are developed or influenced through public and private media. Some recent examples are the currently fashionable low-cholesterol or high-fiber foods. Products designed to fulfill this perceived need charge into the marketplace with built-in momentum. Once the consumer makes the first purchase, the product must, of course, pass all the other requirements of palatability, practicality and economic value. For the product to achieve any meaningful lifetime, it must perform to consumers' expectations, reinforcing their initial perceptions. But first and foremost, consumers must decide whether or not to make the initial purchase.

More and more, contemporary food choices have focused upon health

125

FIGURE 7.1 Freedom of choice.

and quality of life. Status and, more particularly, self-expression, still moderate many of our choices, but they are currently not quite as conspicuous as in the previous two decades. As a result, health and safety aspects, along with environmental considerations, occupy a great deal of the food marketer's time and attention. This has occurred to such an exaggerated extent that it has resulted in fads or fashions rather than rational food decision making. We now consume the high-fiber diet, the oat-fiber diet, the Mediterranean diet, the low-cholesterol diet, the macrobiotic diet and countless other diets in our rush to be healthy and remain young. The diet fads have replaced the perpetual search for the Holy Grail. These various diets are devoured to such an extent it is no surprise that the food-related health problems which are on the rise are a result of over-consumption. It is all a reflection of our rather immature perspectives on food. Simple balance and moderate consumption do not seem to command our attention. The sensible advice given by food and health professionals does not attract the consumer as much as the radical, magic bullet approach of a special diet.

The media have played a central role in this phenomenon. Simple, rational advice or statements do not attract most people. As the media can only survive if they attract and maintain customers, the controversial approaches have been stressed to the virtual exclusion of the simple, sensible, more reasonable ones. Decades of this type of exposure seem to have left most of us incapable of reacting to sound advice. We only respond to shocking, radical approaches. This phenomenon is rather contradictory, because most of us implicitly know that a more balanced approach to health and eating habits makes more sense. But we go along with diet fads anyway.

The natural food sensation has shown spectacular growth over the last two decades. Despite the lack of large-scale industrial support during the initial years of natural foods, they continued to gain popularity because of the consumer perception that natural foods were more nutritious [145]. Once the larger food companies saw there was little use in fighting the trend, natural-type foods became a major thrust in the industry. Even junk foods became natural. The definition of the word *natural* was stretched to such an extent that it bore little resemblance to what most consumers consider natural. The statement, ''Derived from all natural ingredients,'' could mean anything from asphalt to zeaxanthin (a carotenoid pigment extracted from maize), yet most consumers innocently depicted green fields, quietly grazing cows and little gamboling sheep.

Consumers did not picture large stainless steel fermentation vats controlled by technicians in white laboratory coats and knee-high rubber boots running around twisting knobs, reading vapor gauges and issuing orders to ''hose everything down.'' They did not visualize naturally derived, healthy vegetable oils being extracted with hexane, treated with hot alkalis and distilled in cracking columns that would do the petrochemical industry proud. They saw fit, handsome, middle-aged men resting on mountain tops eating healthy foods after vigorous nature hikes. Of course, the image did not indicate that the same men had to climb the mountain in order to get away from the nauseous fumes of the ''natural'' oil refinery.

Consumers are very impressionable. This fact alone has made the marketer king of the hill in the retail food business. The person who can deliver the impression is the individual who, to a large extent, controls public eating habits. This is not a new phenomenon. White bread or refined white flour is a good example. Jonathan Swift, in *Gulliver's Travels*, complained that a simple, unrefined diet was far healthier than the refined foods that were already coming into vogue in the early eighteenth century. During the last century, the milling industry strove towards whiter and whiter flours. Once most of the bran had been removed, flours were milled into very fine particles so that the increased

FIGURE 7.2 Diet fads.

number of reflecting surfaces made the finished product brighter. Then bleaches were added to make the flour still whiter. When all else failed, the bags in which flour was sold were fitted with a very dark inner liner so that when the consumer opened the bag, the visual contrast between the liner and the flour made the product appear even whiter. All this was done without ever asking consumers their opinions on the desirability of whiter flour. (In all the years spent in the flour and baking industries, I never had any consumers complain that they wished to have whiter flour—quite the contrary.) The natural phenomenon changed this, but opened up a new avenue for marketers to follow [146]. Hence natural and ecological products for predisposed consumers.

Marketing which makes use of preconceived ideas or biases is obviously opportunistic, and is far easier and cheaper than educating the consumer. It basically consists of reinforcing a consumer's prejudice. During my time in the baking industry, I developed and patented a "white" bread that had the same digestibility characteristics as whole wheat bread by replacing wheat bran with the hulls of yellow field peas [147,148]. This was in response to the instinctive belief consumers had in the value of whole wheat bread, yet who nevertheless ate white bread. Therefore, a white bread was made that had the same digestibility characteristics as whole wheat bread. The product was even called "Whole White" bread to reinforce this notion [149]. Variations of this high-fiber white bread made with pea hulls are still being sold today in large volumes because it continues to meet a perceived need.

The natural approach has been taken up by both pro- and anti-food irradiation advocates. The pro-food irradiation advocates stress that the technology is a more environmentally friendly way to control foodborne diseases and food spoilage without resorting to harsher treatment or chemicals. Unfortunately, most consumers are simply not aware of the current extent of foodborne diseases or food losses. These two issues do not rate highly as topics of social interest, even though their impact is real and measurable. (In fact, because they are real and measurable, these issues do not attract New Wave advocates, since their main trade is in areas that are not easily quantifiable.) The anti-food irradiation advocates stress nuclear, environment, cancer potential and general distrust of technology which take us away from the natural way of life. It is a mixed-bag message to reinforce a wide range of built-in fears and prejudice. One would expect that this message should have won the debate hands down. This has not been the case because consumers are starting to become as wary of New Wavers as they previously were of scientists.

The anti-food irradiation advocates have repeatedly said consumers would not accept irradiated foods, and have even quoted their own opinion polls to reinforce this view. If they really believed this, then they wouldn't have set up the quasi-military papaya night patrols in California to ensure that consumers would never get a chance to make that decision at a planned test market [150–152]. Despite the disruptive efforts of the National Coalition to Stop Food Irradiation, the final results clearly revealed that consumers preferred irradiated papayas. But when the NCSFI troops finally landed with their picket signs, the tests were suspended.

When the facts are all considered, they do not in any way support the notion that consumers don't want irradiated foods. Regardless of what has been said or written about irradiated foods, consumers have never rejected them. Quite the contrary! Wherever fully labeled irradiated foods have been introduced or test-marketed without interference, consumers have been overwhelmingly in favor of them. All the published surveys which have fully disclosed the methodology used, indicate that consumers will readily accept irradiated foods [153–157]. It was also clear in these surveys that most of the consumers who were not willing to buy irradiated foods initially, changed their views as soon as they

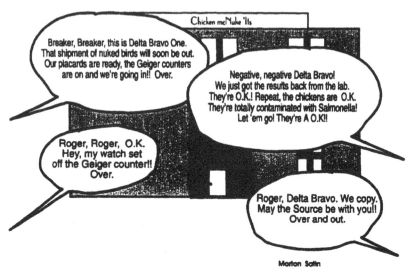

FIGURE 7.3 The poultry patrol.

understood why irradiation would be used, i.e., reduction of disease-causing bacteria, reduction in use of chemicals, effective insect quarantine treatment, etc.

One of the first trial markets was carried out in Florida on September 11, 1986. Four-hundred-eighty cases of mangoes irradiated in Puerto Rico were placed on sale at Laurenzo's Farmer's Market in North Miami Beach [158–160]. Both the vice president and produce manager were not particularly familiar with food irradiation, but presumed there could not be any problems, since they were permitted to be sold by the U.S. Department of Agriculture. In fact, they decided to promote the mangoes by putting up large signs around the store indicating that they had irradiated mangoes for sale. The net result was very successful, with purchasers coming back for more irradiated mangoes after they had tried the first ones.

As soon as the press wrote up the story, Laurenzo's was surrounded by crowds of anti-irradiation protestors. In fact, one of them collapsed in front of the supermarket store and complained he was suffering from radiation sickness. When David Laurenzo informed him that the mangoes were not in the supermarket itself, but rather in the farmer's market across the road, the protestor quickly got up and ran away [161].

Thailand is an excellent example of a place where irradiated foods have been successfully introduced onto the market. A traditional Thai delicacy is called nham. Nham is a sausage made from pork, pork rind and rice, and is fermented for 3–4 days. It differs little from the fermented sausages found in other countries except in its higher moisture content. This increased moisture, unfortunately, supports the growth of undesirable microorganisms and parasites such as *Salmonella, Shigella, Trichinella, Taenia*, and *Entameba*. Despite its fearful reputation, nham is regularly consumed most often as a raw product, since cooking greatly changes its flavor. (Many people abhor the thought of consuming raw food, yet eagerly eat smoked salmon, fresh mayonnaise made with raw eggs, raw ham, clams and oysters.)

The popularity of raw nham, and its public health risk, led the government to sponsor a long-term study into the safety and acceptability of irradiated nham. Multigeneration rat studies using animals fed on normal laboratory feed, feed containing 50% non-irradiated nham, and feed containing 50% irradiated nham were carried out. Cytogenetic (polyploidy), teratogenic (birth defects), blood, longevity, fertility, carcinogenicity, and pathology tests were carried out on all generations. The results clearly indicated the safety and wholesomeness of the irradiated nham [162]. Once the public health value of irradiating nham was established, it was allowed to be sold on the market. This is currently being done and the market for this product is growing rapidly. The product packaging not only indicates that the product has been irradiated, but also the reason why it has been irradiated. Consumers are thus made aware of the dangers involved in the consumption of untreated nham.

In most other countries, however, consumers have not been given the option to choose irradiated foods, despite the known benefits. In such a situation, the only means of determining the acceptability of irradiated foods is through test marketing. Where these have been carried out, consumers have chosen irradiated foods, even when their price had been arbitrarily placed higher than untreated products.

A good example of such a trial was that carried out in 1987 in France on strawberries. In France, the word *irradiation* is closely related to nuclear fallout, and has very negative connotations. The government therefore agreed to the use of the correct term *ionization* as a more meaningful alternate. Thus the irradiated food used in the test market was labeled *Protégé par Ionization* (protected through ionization).

In France, consumers are rather particular about their food. Quality and freshness are very important. Given the choice, consumers would prefer freshly harvested, non-processed foods even though they know it is not possible, given the growth of urbanization and the complexities of the distribution system. Strawberries are a springtime favorite, but their ready availability through nationwide distribution has dictated a technology which has resulted in a lower overall eating quality. Many of us have had the pleasure of going out to the country and tasting ripe, freshly picked strawberries and therefore know what they should taste like. Unfortunately, in order to make the same fruit available throughout all the urban areas, strawberries must be picked fairly green, packaged and put into the distribution chain. Full ripening takes place during distribution.

Strawberries are fragile and easily subject to molds. How many of us have selected those familiar green plastic containers covered with clear film only to find some moldy strawberries under the beautiful ones on top? Thus, if irradiation allowed strawberries to be picked at a riper stage, their color, flavor and shelf life could clearly benefit. Consumers react very positively to this, particularly when they are actually presented the product in question [163].

Irradiated strawberries have lower levels of microbial contamination and are thus less susceptible to spoilage during distribution. Therefore, in the French market trial, they were picked ripe when their color and aroma were at peak. In addition, the containers they were packaged in were clear, rather than opaque green plastic, to ensure that consumers could see that there were no moldy strawberries in the bottom layers.

The irradiated strawberries were sold side by side to unirradiated ones, but at a price 30% above the untreated ones. They were fully labeled to ensure that consumers knew they were treated and were guaranteed to retain their freshness four full days. Despite the very significant difference in price, the turnover of irradiated strawberries was equal to the untreated ones. There was a telephone survey in 1987 on the willingness of consumers to pay more for the potential benefits of irradiated fresh foods. The results indicated that more than 50% of the consumers who would purchase treated strawberries would be willing to pay more for the extra length of freshness retention that irradiation provides [157]. Studies also indicated that consumers would be willing to pay more for foods that provide a greater degree of safety [164]. This should not be surprising at all, since we never question the far higher prices some people pay for the allegedly organic or natural foods they feel may be safer or better for them.

Another example was the previously cited test marketing of irradiated papayas in California. As a result of the ban on the use of ethylene dibromide (EDB), the Hawaiian papaya industry uses the double hot-water dip technique to ensure fruit disinfestation. This method is not ideal, since it can damage the fruit. The fruit must therefore be picked

green in order to withstand the treatment. Irradiation is an ideal disinfestation procedure because it is very effective and has no negative
impact on fruit quality. Market tests were carried out in two California
supermarkets in 1987 in order to determine consumer response to
irradiated papayas [165].

During the market trials, the irradiated papayas and the regular double
hot-water dipped papayas were sold side-by-side. The papayas treated
with irradiation were clearly labeled as such. The net results indicated
that the irradiated papayas outsold the double hot-water dipped ones at
a ratio of better than ten to one. No wonder the National Coalition to
Stop Food Irradiation activists tried to disrupt the tests.

Why then, have the actual market results been so different from the
dire predictions of anti-food irradiation advocates? Why do they insist
that consumers will never accept irradiated foods? The first reason is
that they are convinced no amount of evidence can refute their own
judgments. This is symptomatic of the moral imperative, and again
reflects the tyranny of the self-righteous. The other reason is the results
of their own surveys. Why have the results of their polls differed so much
from actual, proven results in the marketplace? This is most likely
because, when making actual food choices, consumers must see tangible
evidence of the products and their benefits, rather than simply reacting
to leading verbal or written inquiries regarding their potential acceptability. This is a critical point, because anti-food irradiation advocates
fear that such foods will be readily accepted when available in the
marketplace. Thus, their strategy is to do everything possible to prevent,
or delay for the longest possible time, the introduction of irradiated foods
to the marketplace.

In summary, the only evidence of market trials we can legitimately
refer to clearly indicates that properly labeled, irradiated foods are
acceptable to a very large segment of the consuming public and may, in
some cases, be more in demand than those foods currently available.

Why then are irradiated foods not available?

Who Permits Freedom of Choice for Consumers?

Government's role

Although food production and marketing in most western countries is
in the hands of the private sector, the government, or public sector does

play an active role in the development of policies and implementation of legislation that affect the quality, price and availability of foods consumers eat. Import regulations and trade barriers determine the price or availability of foreign products. Production subsidies (including distribution incentives) affect consumer food prices. Finally, the standards of quality and safety of foods available to consumers are largely controlled by the government. The agencies responsible for the latter function are also often responsible for the provision of nutrition and food safety information to consumers.

The policies involved in food pricing and availability demonstrate a complex relationship between the producers, the government and consumers. Despite the increasing influence of consumer advocacy groups, however, consumers continue to have relatively little impact upon government policy in these two areas. Policy development in the area of food pricing and availability remains largely an issue between the producers and government. The only obvious exceptions have been government programs to subsidize food purchases for specific groups of people at risk in certain countries (food stamps for the poor, or food discounts for the elderly).

The main impact consumers have had on food policy development has been in the area of food quality, and the availability of information on most foods. (Notable exceptions to this are restaurant and institutional foods, as well as alcoholic beverages.) Government policy on food quality usually relates to food standards and safety. Food standards can be as simple as the grading of fruit according to size and color, or as complex as the standards for a product, such as enriched white bread. These include standards for enriched white flour (complete with vitamin and mineral supplementation, bleaches, maturing agents, etc.), maximum levels of addition of mold inhibitors, emulsifiers, fats, water, milk or whey powder, dough improvers and other ingredients depending upon the individual country.

All the ingredients, additives and processes used in the preparation of foods must be considered safe by the government. Ingredients that have been consumed for generations are sometimes considered safe simply because no ill effects have ever been specifically attributed to them. The comprehensive safety tests are waived for materials "generally recognized as safe" (GRAS). The logic behind this approach is somewhat flawed, but the practical considerations of costly long-term testing of traditionally consumed food ingredients have prevailed. In fact, very few GRAS substances have proven to be problems. (The weakness in this system is the criteria used to define what has been traditionally consumed. For instance, if a material has been used as a whitening or dispersing agent in tablets such as aspirins or vitamins for years, should it be generally recognized as safe for use in foods where it will be consumed in much higher quantities?)

This issue rarely involves food processes. Most food processes are physical and do not add anything to a food; they simply change the food physically and chemically through the formation of lytic products. Freezing, microwaving, canning, heating and irradiation are examples of such processes. Even age-old heat processes produce many questionable by-products, such as lysino-alanine, a double amino acid which is formed when proteins are heated, particularly under alkaline conditions. Lysino-alanine has been implicated in the formation of enlarged kidney cells, but the foods are still considered safe because they have been routinely consumed since the time fire was discovered. The only way to avoid lysino-alanine, is to stop boiling or pasteurizing milk, cooking meat and, for Mexicans, giving up their tortillas. Ideally, everything could be consumed cold, but again there is the question of foodborne diseases. Therefore decisions have to be taken, and those decisions must be based on real evidence and experience, not irrational projections.

Smoking and bottom broiling are exceptional cases, since they are processes which do add substances to foods, namely smoke components and pyrolysis products. Pyrolysis products in broiling result from fat dripping onto the coals where they are transformed at very high heat, vaporized and rise up to be deposited onto the product being broiled.

In the development of standards or safety criteria for foods, consumers have little direct input in policy development. Standards are normally evolved through negotiations between producers or processors and the government. Practical production considerations, combined with scientifically derived safety data are the key criteria employed to arrive at these standards. From the dawn of time, consumers have been continually subjected to toxins of all types in their food and in the environment. Our biological mechanisms have evolved to handle them quite efficiently. It is only when the level of these toxins exceeds our ability to handle them, do we face problems. The setting of safety limits thus takes into account the tolerance we have towards specific toxins – with a considerable margin of safety built in.

Toxicologists and microbiologists are the scientists who have been trained to provide us the information upon which to make decisions on food safety standards. Consumers generally have little to say regarding these standards. Neither should politicians, businessmen or New Wavers. In everyone's interests, these determinations should be done objectively by trained professionals and without any outside interference. It is a complicated enough process as it is. For example, when establishing microbiological criteria by assessing the risks for certain foods, three specific questions are often asked:

- Does the product contain sensitive ingredients which will make it more susceptible to spoilage or pathogenic bacteria?
- Does the production process have a stage or process in which offensive organisms are destroyed?
- Does the distribution, retailing and consumer handling of the product allow it to become harmful?

The process of establishing such standards is sensitive to all the realities of food production and is a complicated affair. It is far more complicated than simply setting wishful standards that are unreal and unachievable. The various forces that work on this system in order to influence it and remove its objectivity are the greatest dangers to food safety. Politicians, lobbyists (environmental, industrial, consumer) and the media are perhaps the greatest danger to the development of objective standards. Entrusting this process to professionals in government has been one of democracy's great challenges. It has not always worked. Public opinion, influenced by special interests, prevented the enactment of pasteurization legislation in some European countries for decades. This delay resulted in millions of cases of needless illness and most probably thousands and thousands of needless deaths and cripplings. The system of setting honest and rational standards is always under attack from external influences, but usually effective legislation is eventually arrived at.

If certain consumers do not agree with or accept these standards, they make their feelings known in the marketplace. They simply do not buy the products they don't believe in.

Additives are a good case in point. The permitted levels of various additives allowed in foods are safe. Yet, a number of consumers prefer to eat products without additives. In some cases, this means products with significantly decreased shelf life, or poor appearance. Nevertheless, certain consumers prefer these products to those containing additives. Their feelings have been expressed in the marketplace through the growth in sales of organic or natural foods.

There is a growing concern about the possible environmental effects of the chemicals which are commonly used to control sprouting in the potatoes we eat. This may one day be rejected by consumers in favor of irradiation, which has no negative environmental effects. The point is that once the scientists determine that a product or process is safe and useful, consumers are normally permitted by the government to exercise a free choice in the marketplace, and thus express their likes and dislikes.

Where consumers and consumer advocacy groups have had a very significant impact on policy development is in the area of consumer information. This area is of extreme importance, because it has contributed to upgrading the basic concept of a free choice to a free and informed choice, as long as the information is honest, objective and untainted with prejudice.

Influences on the quality of the foods available on the market originate from producers, processors, retailers, nutritionists, and to a lesser extent, consumer groups and the media, all of whom contribute in one way or another to the food chain and consumers' perceptions of food.

The entire food chain from producer to consumer is rather complex, with key policy decisions and judgments made at various stages throughout the continuous sequence. If we take the standard food marketing system, policy decisions on the various aspects of quality are made at the producer, processor and retail levels and are influenced by their corresponding concerns.

Producers and processors

Once policy decisions have been made, it is the producers or processors who next influence the quality of foods available to consumers. Some processors feel that consumer prices are a key consideration, and place a priority on bringing foods to the consumer at the lowest possible price. Others believe they can be more profitable by selling more expensive products through the use of costly ingredients, fancier packaging and image marketing.

The principal areas of new product and technology concern for food processors are as follows.

- usefulness of the technology, costs and the ability to improve the processor's competitive position
- health and safety considerations
- government regulations, including labeling
- public image, consumer perceptions and consumer acceptability

Another critical consideration is the capital investment which is in place for conventional technology. For example, will irradiation make those facilities worthless? Will conversion to food irradiation be an open admission that the foods currently offered on the market could and should be improved?

Of the above considerations, the most cogent for the processor are competitiveness and consumer acceptability. Negative results on both these issues would indicate a technology which is not viable. Yet, quite the opposite is the case with food irradiation. However, the perception of food irradiation's image and the threat of public vilification has served as a form of extortion, and influenced many to shy away from the technology. For the most part, they have not done so on the basis of scientific or technical considerations.

Retailers

Finally, retailers make the decisions as to which foods will be offered to the public. The retailers also decide the priority (or shelf space) they wish to give to various products. Profitability is generally the key consideration, but direct contact with the public makes the retailer very sensitive to consumer advocacy groups and the media. Fear of public vilification, rather than technical, nutritional or market considerations, has moderated the actions of many supermarkets. This is unfortunate for two reasons. The first is that they have not considered the evidence that every legitimate consumer market trial has shown – that consumers find irradiated foods very acceptable [233]. The second is that they have allowed a small number of vocal radicals to cower them into treating their consumers with disdain and ignorance. Retailers, more than anyone, know the incidence of *Salmonella* in chickens. Most retailers know that fruit quality and variety could be vastly improved. Large supermarket chains often have an in-house staff of scientists employed to give them professional advice. Yet, many supermarket giants have pandered to a few New Wavers, and have left their own consumers with

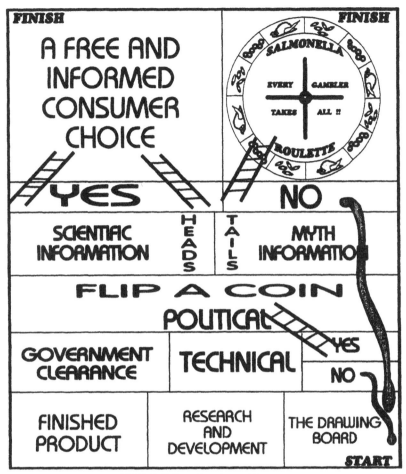

FIGURE 7.4 Sneaks and leaders.

the short end of the stick—foods that continue to contribute to consumer illness and waste. So much for courage and for public responsibility.

What picture does this whole system leave us with?

The government views

Professional civil service (ministries of Food and Drug, Health and Welfare, etc.)

Food irradiation is safe. There are some very beneficial uses for food irradiation in the prevention of foodborne diseases and in the preservation of foods in a fresher condition. The technology poses less environ-mental risk than alternative methods for storage and disinfestation. (These internal scientific evaluations can be found in the files of all governments.)

When instructed by the country's legislators, the regulatory system will rapidly accept the production and sale of irradiated foods to the consumer.

Politicians

Food irradiation is a touchy issue which has been connected to the nuclear debate. It is an emotive issue and has a potential to polarize constituents. Even though the scientists say it is safe and beneficial, most politicians feel it is better to avoid the issue than to support it. As long as there is no public pressure to allow its introduction, the politicians will leave it alone. If anything, a negative stance has generally been taken to appear consistent with publicly stated positions on nuclear issues. They have also avoided, where possible, criticisms of those traditional technologies that could be improved through the use of food irradiation (i.e., production of *Salmonella*-reduced poultry).

The food industry view

Scientists and technicians

The technology is safe. It may become more competitive with con-ventional technologies, and, particularly when it becomes popular, there will be incentives for industry to do more application research. The current applications are very limited because most of the previous research carried out by government and academic institutions was primarily directed at establishing the safety of the process.

Executives

As long as there is no one else using the technology, it is generally felt there is no need to be the leader in a controversial field. Although there is no question as to the safety of the process, they believe that it is best not to take any hard position on the technology since no immediate benefit can accrue from it. On the contrary, food irradiation is capital intensive and may replace some conventional technologies where significant capital funds have already been invested. The poultry industry now sells every chicken it can produce, regardless of its pathogen content. Neither the fast food chains nor the supermarkets demand low pathogen levels – why rock the boat?

The food retailer's view

Scientists, technicians, and distribution professionals

This technology could be a real advantage to business. Currently, there is a very limited choice in the ability to purchase goods that are free from certain foodborne disease microorganisms, yet the retail industry bears a large part of the legal liability associated with food poisoning. This technology could significantly reduce the risk to consumers, and the retailers' consequent liability.

Another great advantage is associated with the perishability of fruits and vegetables. The market for exotic tropical fruits or vegetables is largely limited by their cost and availability. Food irradiation can significantly improve their market position and make all produce cheaper because of reduced wastage. The same situation applies to fish and seafood. Price and market access are essentially a technical problem which the food irradiation process is ideally suited to address.

Executives

Public relations is a key to the retail business. This is an issue where the anti-food irradiationists have been much more vocal than the scientists. Food irradiation has been made into an emotive issue. As long as the press keeps quoting the anti-food irradiation lobby more than the scientists, then the food retailing industry will have a tendency to state that it will not sell irradiated foods. The retail executives want to ride the right side of the public relations scale. This is not based upon any scientific judgment, but rather the wish to be perceived as consumer conscious. Considering the growing loss of credibility that New Wave advocates are experiencing as a result of their tactics, this view of retail executives may change dramatically in the future.

(The anti-food irradiationists have used some rather intimidating tactics such as getting companies to make public statements for or against the use of food irradiation. The environment in which this question is posed is emotional and political, but neither scientific nor health-related. People who are in favor of irradiated foods are painted as anticonsumer villains.)

The politicians and executives are the real decision makers in the final analysis. Understanding their positions and attitudes is critical in any attempt to influence their decisions. It may be worthwhile reviewing some of the pressures that they have faced throughout the food irradiation debate.

Influences upon Decision Makers

The following are a selection of statements or questions that decision makers have faced in the past.

In 1985, the London Food Commission (a consumer activist group), carried out their pilot survey on food irradiation [123]. Some of the particular questions asked of industry were:

> Do you have access to equipment or expertise for testing that would enable you to detect whether food has been irradiated?
>
> [This was a rhetorical question, since it was generally known that no such tests were available at the time.]
>
> In particular, we would welcome your comments on the question of whether food that would fail current hygiene standards at any stage in its handling or processing should be permitted to be irradiated and marketed after irradiation.
>
> [Again, this is a leading inquiry and it is hard to imagine anyone responding that food failing current hygiene standards should be permitted to be sold after any treatment. The implication is that irradiation is an open invitation to abuse on the part of all manufacturers.]

These types of questions are specifically designed to assure a particular response. As it happened, the majority of replies from respondents indicated that they were undecided or had no official policy or comment. The attempt at entrapment was obvious to everyone.

Other public statements or stances on the part of anti-irradiation advocates have been even more explicit.

A four-page flyer issued under the heading of Project Cure, Washington, D.C., states, among other things:

[On page 1]: You see, while no one was watching, the giant food processing industry was somehow able to convince the FDA that nuclear radiation is perfectly safe for use in our food!

However, experimental studies show:

- EVERY rat fed food treated with nuclear radiation in a controlled laboratory test DIED!
- Children in India who were fed irradiated wheat immediately developed BLOOD ABNORMALITIES!
- Mice fed doses of irradiated chicken quickly developed genital cancers!

What's more, literally HUNDREDS of nuclear food processing plants will have to be set up across the nation to "process" our food by shooting it with nuclear radiation.

And each and every one of these plants (which must be built either near a major city or in the middle of an important agricultural area) is a potential disaster which could equal the Chernobyl nuclear incident in Russia.

[On page 2]: This nuclear nightmare MUST be stopped before it comes true!

For the sake of cutting their costs and increasing their profits, the giant food processing industry is willing to endanger the health of every American.

And YOUR government seems to be willing to let them get away with it!

Has the FDA gone crazy?

Remember, EVERY rat fed irradiated food DIED!

[Page 3 of the flyer goes on to say near the bottom]: I cannot win this fight against the multi-billion dollar food processing industry alone.

I MUST have your help in 3 important ways today:

1. Sign and return the enclosed RADIATION WARNING TO CONGRESS petition to me today. Please do not send it directly to your Congressman. When the timing in this lobbying battle is just right for maximum impact, Project CURE will deliver it to your Congressman for you.
2. Sign and mail the enclosed postcards to your two United States Senators.
3. Get out your checkbook right now and make out a check to Project CURE for the most generous emergency contribution you can possibly afford.

[And on page 4 of the flyer]: I have no way of knowing your personal circumstances.

Perhaps you're fortunate enough to be able to afford to make an emergency contribution of $50, $100, $250, $500, or even $1,000 to help Project CURE wage this vital lobbying battle against food irradiation.

Perhaps $15, $25, or $35 is the most you can spare to keep irradiated poison off the shelves of America's supermarkets.

PLEASE mail your two postcards to your Senators TODAY.

And PLEASE try to include the most generous contribution to Project CURE you can afford when you return your signed RADIATION WARNING TO CONGRESS petition to me today.

In another printing entitled *Food Irradiation: To Whom It May Concern*, written by James Marlin Ebert, Ph.D., the director of the Central Texas Chapter of the National Coalition to Stop Food Irradiation, makes dire predictions on the future impact of food irradiation and urges consumers to action. He also asks who has the God-like knowledge to alter something from nature for our betterment. The footnote at the bottom indicates that the printing was donated by the Organic Food Network of Austin, Texas.

One of the most sensational of the anti-irradiation groups is Food and Water Inc., formerly of New York and currently based in rural New England. Food and Water, under the direction of Walter Burnstein, an osteopath from New Jersey, took the lead in opposing the opening of the first U.S. food irradiation plant. Food and Water ran a series of ads which stated that the consumption of irradiated fruits and vegetables "might kill you." Burnstein later admitted that the ad was extreme, but justified it on the basis that they were desperate to stop the technology [166].

Although such claims are outrageous and take no account of the scientific evidence available on the subject, Food and Water received a substantial amount of funding from the Mary Reynolds Babcock Foundation. This funding institution, which may have access to the scientific resources of Reynolds Tobacco to rationally evaluate all the evidence on the matter, chose to list "education" as the reason for their support of Food and Water, Inc. This fact, along with a comprehensive description of Food and Water's origins, methods, funding and tax status are described in a recent article which provides insight into the operations of such organizations [167].

Another group which has solicited financial assistance from the public in their drive "to halt all food irradiation until Congress commissions an independent study by the National Academy of Sciences on the safety of . . . and alternatives to . . . food irradiation," is the Center for Science in the Public Interest (CSPI). This statement of purpose itself is revealing because of the explicit "alternatives" condition. If an independent study has determined that food irradiation is safe (and all independent international technical organizations have done this), why is there the condition on alternatives? If the safety of the process has been assured, it does not seem reasonable to delay the introduction of food irradiation until a functionally equivalent alternative has been elaborated. (This would not have been workable with pasteurization, because, to this day there are no acceptable alternatives to this technology.)

In an undated mailing, CSPI notes that irradiated foods are not radioactive and the process does kill "many of the molds, bugs, and bacteria that grow in foods—assuming they are given a good jolt of Cobalt 60." This gracious concession to the functional value of food irradiation is iterated under the heading "Giving the Devil Its Due." The mailing goes on to list the standard complaints of radiolytic products, vitamin losses and environmental and worker risks. Of course, there is no mention of the equivalent or greater effects resulting from the other technologies currently used.

It is difficult to see how such comments can be called "science in the public interest," particularly when the independent scientific data clearly describes the current levels of foodborne illnesses and the near 100% rate of contamination of poultry with either *Salmonella, Campylobacter* or *Listeria*. What is clear, however, is that requesting money in order to prolong the debate with yet another review is definitely in the interest of CSPI.

The story goes on and on. Mrs. Bloch von Blotnitz, a member of the Green Party in Germany and Rapporteur of the European Parliament's Committee on Environment, Public Health and Consumer Protection stated that food irradiation "makes it easier to deceive consumers." She went on to describe irradiated foods as being an "empty shell." Such has been the quality of consumer representation in this debate. It is little wonder that such an emotive approach tends to polarize people. Most decision makers therefore avoid the issue altogether, or use the opportunity to gain short-term support from the side they feel is winning public opinion. This stance may be understandable, but it is irresponsible because it sidesteps the current problems that food irradiation can effectively address.

Even the well-known International Organization of Consumer Unions (IOCU) has in the past requested a worldwide moratorium on the use of food irradiation. In the face of all the established information available on foodborne diseases, and the judgments made by respected international scientific authorities on the safety and merit of food irradiation, it is a wonder that they based their recommendations on worn-out, hackneyed criticisms that were never valid in the first place. IOCU is mandated to bring a consistent level of high quality to the task of consumer advocacy. Recently IOCU has taken a more independent and unbiased approach to food irradiation, and will hopefully provide their clients with a better and more balanced input in the future.

Positive Consumer Action

Regardless of the issue in question, consumer actions can be taken in two ways, individual and collective. Individual actions can take the form of letter writing, calling or otherwise informing those who make decisions on consumer choices. Collective action through consumer organizations can also take the form of calls or letters, and are often more effective. Consumer organizations are generally in close contact with the media and can thus be very useful in influencing politicians and retailers. It is even possible under proper circumstances to bring issues to court, or to the local ombudsman, in order to gain the attention of decision makers.

Legislation

Foods that contain unacceptable levels of disease-causing microorganisms can make consumers sick. Foodborne diseases are naturally not part of the contract between the buyer and the seller of the food. As a buyer, the consumer has been mistreated, and has recourse against the seller and manufacturer. If the consumer has suffered food poisoning, the pain, discomfort, loss of work and distress that he or she has experienced must be compensated for by the seller and manufacturer.

Effective protection of the consumer under these conditions requires that:

- Legislation is in place making it unlawful to sell food contaminated with unacceptable levels of disease-causing or spoilage microorganisms. (This is not the present situation. Because pathogens can occur naturally on foods such as poultry there is no legal limit on their levels.)

- Consumers can quickly and easily determine if they have suffered food poisoning.
- Consumers must be notified of the monitoring of foods through a public listing of those found to be above the legislated limits, and where they were purchased.

With reference to legislation, it would be worth any consumer's time to contact their government food and health ministry, or department of agriculture and ask precisely what laws are in place to protect consumers from food poisoning. Some of the questions to be legitimately asked are:

- Are there regulatory limits for foods contaminated with disease-causing microorganisms? What are they?
- How frequently are foods tested for disease-causing bacteria? Are they tested at the manufacturers' premises or at the end of their shelf life in the supermarket?
- What are the penalties to manufacturers or retailers if their foods are found to be contaminated with unacceptable levels of disease-causing or spoilage microorganisms?
- If the average incidence of *Salmonella* in poultry is $25-80\%$, why have more manufacturers and retailers not been prosecuted?
- If the combined incidence of *Campylobacter* and *Listeria* is near 100%, why have more manufacturers and retailers not been prosecuted?
- What are the legal alternatives to consumers if they get food poisoning?
- What do government scientists recommend to reduce the incidence of foodborne diseases in foods, and why is there such a high incidence of contaminated foods sold in retail supermarkets?

These questions really have nothing to do with food irradiation per se, but rather with the current state of foods available at the supermarket. If the legislation in place seems less than adequate to protect consumers from foodborne disease, it is simply because an insufficient amount of political pressure has been placed upon the legislators. On the contrary, it is possible that companies wishing to maintain the status quo on microbial legislation in foods have lobbied to delay any changes.

Another, more direct question is:

- Why is there not a choice of contamination-reduced irradiated foods available to consumers? Who is responsible and what are precisely the technical reasons for this, in light of the government belief that the process is safe?

The significant number of immunocompromised people, such as those with AIDS, are a particular case. There are no accurate statistics currently available as to which vectors hold the greatest potential for fatal infections to them. The fact that many victims under intensive care receive irradiated foods is a clear indication that foods are a common reservoir of harmful bacteria — not surprising given the known incidence of pathogens in common foods. The well-organized political influence of these groups may make it possible to mount an effective lobby to influence legislators to tighten the limits on offensive microorganisms in foods, which will benefit one and all.

Consumers should have access to a means of conveniently and rapidly determining if they have food poisoning. Officials of departments of health can be contacted and requested to provide information as to those hospitals or health centers which have the facilities to carry out such tests. It must not be forgotten that a very small fraction of food poisonings are actually reported. Knowing the symptoms can prepare consumers for dealing with infectious episodes that may otherwise lead to serious consequences. If it can be proven that consumers have been food poisoned, they will have a greater opportunity to be recompensed for the loss and unfortunate experience they have suffered.

An official and public notification of products found to be contaminated with pathogenic organisms can serve several purposes. In the first instance, it will make consumers more aware of the ongoing incidence of food contamination. It will also warn consumers to be cautious of products that repeatedly exceed allowable limits for disease-causing microorganisms. Perhaps, most importantly, it will elevate the issue of foodborne diseases to the priority level on which it realistically deserves to be. The importance of foodborne disease to consumers, politicians, and food businessmen appears to be far lower than to health professionals and food scientists. A public notification may also sensitize consumers to the need of food choices that offer greater safety. This need not be a witch hunt with wild headlines and names. That is a terrorist tactic reserved for the New Wavers. But such a notification should be visible to all consumers (not just those that belong to consumer groups), and should serve to provide useful, not sensational, information.

Market Choices

In the food chain from production to consumption, the two key factors are the manufacturer/producer and the retailer. These can be approached with the following questions:

- Given the published incidence of contamination of various foods with pathogenic bacteria, what have they done to lower the risk to their clients?
- If the government and internationally recognized scientists say that irradiated foods are safe, and serve to reduce the risk of foodborne diseases to consumers, what is obstructing their availability on the market?
- On what technical basis, if any, will the decision be made to provide or deny the option of irradiated foods to consumers?

Consumer Information

Consumers have the right to demand the best quality information from their consumer organization or from the local press. When the information concerns the food a consumer eats, it is not a subject for polemics, or moral opinions. It is an issue which deserves to be addressed by facts and frank, qualified professionals. Therefore, some of the obvious questions to ask are:

- What is being done to improve the quality of our foods from the standpoint of health risks?
- What is the current view of public health professionals on the use of food irradiation as a means of improving the hygienic quality of our foods?
- What are the published comparisons of vitamin losses when irradiation is used as a quarantine treatment instead of conventional treatments, such as long-term cold storage?
- What is the stand on the use of food irradiation as a means to reduce food-related health hazards, and to improve the storage life of foods? Upon what references is this opinion based?
- What is the view on labeling of irradiated foods, and what is it based upon?

Consumers have a right to expect much from their advocates. They should not be satisfied with entrenched positions which are not based on legitimate evidence. In a recent press briefing on the subject of food irradiation, a representative of a well-known consumer organization stated that she would only recommend cold storage as a method of sprout inhibition for potatoes, even though there was no evidence of safety problems with irradiation. A participant from Poland indicated that due to their high level of consumption, potatoes were usually purchased in large bags (25–50 kg) in his country. He added that, aside from being expensive, cold storage was not convenient because potatoes started to sprout soon after they were brought home. Therefore, consumers would have to pay more for potatoes, because they would have to buy them in smaller bags (5–10 kg). Upon hearing this, the consumer representative conceded that in the case of Poland, food irradiation could legitimately be used.

This type of representation should be unacceptable to consumers. If food irradiation is safe in Poland, it is safe in every country. More importantly, if it is unsafe in one country, it should not be used anywhere. It is unacceptable for a consumer representative to demand that consumers accommodate a less practical technology (in this case, cold storage), for no valid reason. If irradiation is a safe method for controlling sprouting in potatoes, why should chemicals be used, or why should all consumers be forced to buy them in smaller, more expensive packaging? This is not qualified advice. (The consumer advocate in question has since left the field of food and is now an environmental activist.)

The standard statement ''We are not sure about this,'' should not be accepted without question. If consumers wanted to be unsure about something, they could be so without having to pay the fees required to join a consumer organization! Consumers join because they want to be sure they get the most qualified information. Consumers must demand, and accept no less than, fully qualified opinions, based on fact. Consumer organizations must, in turn, seek the most qualified sources of information and not be bound by bias, prejudice and the transparent self-interest which inevitably results in the tiresome requests for fees or donations to carry on nonexistent battles. The same predictable styles, arguments and tactics have been seen no matter where in the world they appeared to originate. Consumers deserve more from their advisors and representatives than an inept attempt at manipulation of public opinion in order to secure funding for a cause. Perhaps creative new approaches in the future will lead to more substantial and stable sources of funding for consumer advocacy groups, so they can focus on the task of professional advice and representation, rather than the constant search for a means to support their efforts.

The Media

It is the task of the media to provide factual information. If consumers feel the information they are getting is biased, misleading or simply untrue, then they should write to the media management (letters to editors, station managers, etc.). They may also threaten to stop buying the papers and the products that are advertised in them. If consumers don't react to poor media performance, then they really don't deserve anything better. In the past two decades, the consumer movement made the food industry sit up and take note of their demands for better information. As a result, products are now more consumer-sensitive, and food labels provide more information than ever before. Getting this to happen required action on the part of consumers. The same holds true for the media. If consumers feel the information they need should be more informative, or more solidly based, then they must say so.

The media are educators. Although many journalists don't like to assume this responsibility, others concede that this is true. When a layperson tells others not to eat frozen bread because it could cause cancer, this information is learned from the public media. The media have effectively replaced schools as the lay public's formal access to knowledge. Unfortunately, it is both unfair and unrealistic to demand that all journalists, who are most often under severe deadline pressures, perform as formal educators do.

Notwithstanding the current vogue of presenting the news as a form of public entertainment, most journalists make a considerable effort to provide all the facts. However, the ready availability of New Wavers, combined with the relative inaccessibility of scientific authorities, tends to bias coverage somewhat. Hot-blooded New Wavers are usually far more animated and interesting to speak to than boring scientists. This may not excuse the reporting of false claims and exaggerations, but it does make them more understandable. A good deal of the onus must therefore be assumed by the scientific community, which makes it particularly important for scientists to be readily available and competent to discuss matters with the media. If these people are difficult to access, or are incapable of providing information in an interesting or engaging fashion, then one can hardly blame the media for looking elsewhere for opinions on the issues. Scientists cannot hold themselves aloof from the public on the one hand, and then wonder why no one understands them on the other. Scientists who refuse to be part of the solution end up being part of the problem.

CHAPTER 8

Irradiation and the Food Industry

Throughout the debate on food irradiation, the food industry has remained conspicuously silent. In fact, with few exceptions, even the studies utilizing predictive modeling systems for microbial survival and growth in various foods neglected to include irradiation as a possible treatment [235,236]. This is not particularly surprising since there was neither legislative pressure nor a public outcry demanding improvements in the hygienic state of the food supply. Indeed, the illusion of a food system *par excellence* was perpetuated by everyone in the industry and many in government itself. The time-worn expression, "If it ain't broke, don't fix it!" — the perfect excuse for doing nothing — was the going byword. The few public statements that did surface were made by certain retailers, indicating that they would not carry irradiated foods, because their customers didn't want them. This outlook may have been based on a vocal minority of consumers, since all the available technical literature revealed the high consumer acceptability of irradiated foods wherever controlled market trials were carried out. More likely, this negative stance towards irradiation was little more than pandering to the vocal anti-food irradiation advocates who, through the aid of the press, appeared to have gained the upper hand in public opinion.

This situation began to change dramatically after two pivotal events, which occurred within a year of each other. The first was a prime-time national U.S. network television documentary, which showed the opposing forces involved in the food irradiation debate leading up to the opening of the first dedicated food irradiation plant in Florida. The primary anti-food irradiation advocates, Food and Water, Inc., were

153

shown to have clearly and willfully misinterpreted the available scientific literature and to have greatly exaggerated negative claims, based upon internationally discredited scientific studies. More importantly, a national audience of adult viewers was given the opportunity to see the central characters and *modus operandi* of a fringe group purporting to act in the interests of the public at large. Food and Water fared so poorly on this show that they later requested an investigation into it, claiming there was evidence that the show's producer had a history of discouraging all stories that portrayed nuclear energy in a negative light [237,238]. Shortly thereafter, another stinging criticism of Food and Water appeared in the *Wall Street Journal* [239]. This national exposure to the food irradiation debate rekindled the interest of the public press who took up the subject once again, but with a far more positive attitude towards the technology than they had in the past.

The second event was the Jack-in-the-Box food poisoning outbreak resulting from the presence of *E. coli* O157:H7 in hamburgers. Although it was certainly not the largest food poisoning outbreak of the last decade, a number of factors combined to make the incident particularly noteworthy. The tragic deaths of small children, linked with a nationally known fast-food chain that specialized in America's most popular fast food, thrust the issue of food safety directly onto the front pages of newspapers and made it a feature topic in some of the most widely read magazines and on television shows [240−244].

The "ECO157" incident set into motion a series of government actions designed to address the issue of improved food safety through various interventions such as increased inspection, training, monitoring and legislation [245−247]. The presence of "naturally occurring" pathogens was no longer tolerated with the same empathy as in the past. The medical profession began to speak out in favor of food irradiation as did various state and national agriculture authorities [248−252]. Even the food industry, so long reluctant to get involved in the debate, began to speak out in favor of the potential benefits of irradiation [253−255]. Led by the American Meat Institute (AMI), public statements were issued acknowledging the need to address persistent food industry problems with the best technologies available [256−259].

The AMI had a large poll conducted by the Gallup Organization in order to estimate consumers' interest and intent in buying irradiated foods. The net result indicated that at least half of consumers would purchase irradiated food once they were aware of its benefits [260]. A survey of top food executives indicated that a majority felt that industry should be funding more research on irradiation and that most consumers would be willing to pay more for pathogen-free products [261]. The chief executives of two of the largest meat and poultry corporations in the U.S. came out clearly in favor of the use of irradiation [262,263]. Some of the most well-known and respected legal and business advisory services portrayed the process positively [264,265], and even the Pope gave his blessing to its application [266].

This shift in public opinion towards food irradiation indicated that "something *was* broke" in the food system and definitely needed fixing. If it is believed that public opinion foreshadows future consumer trends, it is imperative that industry now consider food irradiation more seriously in light of the projected demands for a safer, more hygienic and more varied food supply.

A Different Attitude towards Food Irradiation

With few exceptions, the food industry's traditional view towards irradiation has been ambivalent at best. This is quite understandable since the technology has never been considered to be "opportunity-rich." Although food irradiation was able to cope successfully with several major practical difficulties such as pathogen contamination, insect infestation and sprouting, these were not the sorts of problems that the food industry wished to bring to the attention of the public. Furthermore, the perceived association of this technology with the nuclear debate added a new complication to the marketing of foods to a public that was growing increasingly suspicious of modern technology. All things considered, it was generally felt best to stay clear of food irradiation and the stubborn problems that it highlighted.

Now that the public has become far more aware of the need to improve the hygienic quality of certain foods, the industry's attitude towards food irradiation should follow suit. The technology can no longer be perceived simply as a remedial tool exclusively suited to deal with problems of pathogens or disinfestation. It must be carefully considered for *all* the functional advantages it has to offer. Food irradiation can provide solid foods the same protection from foodborne pathogens that pasteurization does to liquids and is also the most effective method for a wide range of residue-free disinfestation and sprout-control measures. However, from the food industry's point of view, the most profound aspect of the technology is that it is able to do this without significantly altering the original character of the food. When various meat or poultry products have been irradiated, the resulting product is still raw and the consumer has full freedom to prepare foods in any manner whatsoever. The same goes for seafood and shellfish.

When food irradiation is considered objectively, free from politics and the rhetoric of the nuclear debate, it is readily apparent that it has the potential to provide a quantum leap forward in bringing better quality, longer lasting and more convenient foods to the consumer. This should not be particularly surprising if a simple parallel is made to the modern use of pasteurization, which is no longer confined to the processing of milk, but is also applied to a wide range of other liquids to provide better hygiene and longer shelf life. Despite mildly noticeable changes in taste and texture, ultra high temperature (UHT) treated products such as milk, cream and juice continue to have very large and loyal markets (particularly in Europe), simply because of the convenience of extended ambient shelf life. It takes little imagination to see how cold sterilization of solids and liquids through irradiation could provide shelf stable products with levels of acceptability and convenience never dreamed of previously.

Thus, food irradiation can no longer be considered merely as a palliative—a technology to alleviate problems that the food industry would rather not talk about. Food irradiation must be regarded as a technology particularly abundant with a functional potential that can be translated into products consumers will desire. It is a technology ideally suited to the continually growing demand for longer shelf life foods. It is a technology ideally suited to the demands of the future.

Risk versus Benefit

Since irradiation is nothing more than a technology, any appraisal of its practical value to the food industry, now and in the future, must follow the same analytical process taken with the consideration of any other technology. Routinely, such an exercise starts with an economic feasibility analysis. This requires a careful weighing of the potential benefits of using a technology against the costs associated with its application.

Unfortunately, the rhetoric and controversy that have traditionally been associated with food irradiation requires that a preliminary judgment first be made on whether to seriously consider this technology at all. This implies a careful examination of all the key issues of concern on both the negative (liability) and positive (asset) sides of the food irradiation balance sheet as follows:

Food Irradiation Balance Sheet

Liabilities	Assets
(1) Outright rejection of irradiated food products by retailers and consumers	(1) Increased confidence in products and product line, resulting from improved hygiene
(2) Negative spillover effect on other products in the company line	(2) Increased overall store sales, because of irradiated food availability
(3) Customers may boycott retailer or manufacturer	(3) Reduced risk of recalls, loss of reputation, loss of sales and costs of litigation and insurance resulting from foodborne disease outbreaks
	(4) Reduced "shrink" in food chain due to better hygiene and improved shelf life
	(5) Greatly improved opportunities for international trade
	(6) Reduced government inspection
	(7) Increased potential for convenient new consumer products

Irradiated Food Liabilities

Rejection of irradiated food

On the negative side of the balance sheet, the concerns are all based upon the widely held belief that consumers will reject irradiated foods when they are offered on the retail market. This misperception results from both the false predictions and the intimidating actions of the vocal anti-irradiation activists. Their threats to boycott or disrupt retail sales, combined with warnings of mail-in campaigns, letters to Congress, etc., have constituted a real nuisance to decision makers [239]. They have also engineered enough negative press in the past to make it appear that consumers will definitely reject irradiated foods. In actual fact, they have proven to be rather ineffective and almost comical in their efforts to disrupt retail sales. Television and newspaper articles have revealed their threats to be empty. For example, when activists threatened to totally disrupt the marketing of strawberries at Laurenzo's, few showed up and store sales were said to be very brisk. When irradiated poultry was introduced at Carrot Top Inc. and Laurenzo's Supermarkets in September 1993, sales were excellent and the atmosphere was "business as usual" [267].

The mistaken premise that irradiated foods will be rejected by consumers is also based upon a number of surveys carried out to determine how consumers viewed irradiation in comparison with other food-related issues (additives, pesticides, etc.). These surveys were generally carried out in an atmosphere of abstract polemics, that is, one in which the end-results of technologies were discussed, debated and judged without the benefit of the participants ever being exposed to the actual material products. Most professionals agree that such results must be treated with a certain caution regarding any conclusions to be drawn on actual consumer decisions at the retail point of sale. The most evident conclusion resulting from these surveys was that consumers would be significantly more interested and positive towards irradiated foods once they knew the benefits involved.

The widely held misperception that consumers will not accept irradiation or irradiated foods has always been based on subjective data and testimony, none of which related to genuine in-store behavior or market trials. When verifiable retail market trials, product introductions and ongoing sales are brought into the picture, the situation changes dramatically. Without exception, whenever fully labeled irradiated foods have been offered to consumers, they have been a dramatic success. This has been demonstrated where the chief advantage for the consumer has been better appearance or shelf life and where the main benefit was a higher level of hygiene in the final product. In other words, consumers appear to consider irradiated products more acceptable for both quality and health reasons. There has not been a single incident where irradiated foods, at the retail level, have been rejected by the majority of *paying* consumers! This is an extraordinary record that fully contradicts any notion that consumers may reject irradiated foods on the marketplace.

Negative spillover

The second point on the liability side, that is, a possible negative spillover effect on other products in a company's line could only hold if consumers reject irradiated foods in the first instance. On the other hand, if irradiated foods are found to be acceptable, there is no reason to assume that there will be negative effects on other products within a company's line. Quite the contrary, a positive consumer perception of higher quality in one product line can spill over to other company products.

Consumer boycotts

The recent introductions of irradiated foods onto the retail markets have taken place without any consumer boycotts. Considering the growing consumer concern over issues such as foodborne disease, this is not very surprising. The growing support of food irradiation by the public press will make it increasingly difficult for anti-irradiation advocates to generate the support necessary to mount any significant boycott attempts. Based on all the available evidence, the fears of such boycotts are largely unfounded and only serve to handicap rational business judgement regarding the marketing of irradiated foods.

Irradiated Food Advantages

Increased confidence in irradiated foods

On the affirmative side of the balance sheet, a number of benefits can be seen. In the United Kingdom, for example, the number of food poisoning notifications has increased fivefold in the last decade [268]. Increased consumer confidence in products resulting from increased hygienic quality is thus a very positive consideration. This is particularly so for certain food categories such as raw poultry, meats and marine products, which are coming under increasing criticism and scrutiny.

Increased store sales

In the retail industry, attracting consumers is the biggest single preoccupation. Full-page flyers, stamp give-aways and loss-leader promotions are examples of the degree of activity devoted to bringing consumers into the store. Increased confidence and interest in products that benefit from improved hygiene and safety will attract consumers to those retail supermarkets that offer them. As the public concern with foodborne diseases increases, consumers will be more attracted to those stores that provide foods claiming to reduce these risks. The latest commercial advertisements for household cleaners specifically state that they reduce the risk of cross-contamination in the kitchen resulting from pathogens brought home with raw poultry and meat products. These full-color ads, describing cross-contamination in simple language or visually depicting a supermarket chicken leaking juices onto a kitchen counter, will reinforce the problem of foodborne diseases in the minds of consumers. Over time, this is expected to result in an increased demand for safer products.

Reduced risks associated with foodborne diseases

On the operational side, the advantages of irradiated products are quite clear. Products with greater hygienic quality will significantly reduce the risk of recalls, loss of reputation, loss of sales, interruption of business [269] and all other costs associated with legal proceedings arising out of incidents of foodborne disease outbreaks. (A recent example is the recall of 3,000 metric tons of snack food products seasoned with *Salmonella*-contaminated paprika powder by the Bahlsen Biscuit company, based in Hannover, Germany. The recall costs alone were estimated to exceed $25 million [270].) Irradiated products will also reduce the costs of insurance required to protect against such losses. This latter point is a growing concern, as evidenced by the number of requests for food poisoning liability insurance that have been registered after the Jack-in-the-Box incident [271].

In an excellent paper recently presented at the conference *Food Irradiation: Issues and Challenges* [272], Mark Robeck, a lawyer with Baker and Botts of Austin, Texas, clearly and methodically portrayed the legal scenario of liabilities on all sides of the food irradiation issue.

Among the statutes referred to, the "Restatement (Second) of the Law of Torts" published in 1965 by the American Law Institute articulates the doctrine of strict liability:

1. One who sells any product in a defective condition unreasonably dangerous to the user or consumer or to his property is subject to liability for physical harm thereby caused to the ultimate user or consumer, or to his property, if
   a. the seller is engaged in the business of selling such a product, and
   b. it is expected to and does reach the user or consumer without substantial change in the condition in which it is sold.

2. The rule stated in Subsection (1) applies although
   a. the seller has exercised all possible care in the preparation and sales of his product, and
   b. the user or consumer has not bought the product from or entered into any contractual relation with the seller.

His analysis of the law, combined with the overwhelming weight of credible science leads to certain inescapable conclusions. A commercial food provider has to be far more concerned about claims for injuries resulting from products contaminated with pathogenic organisms than from any possible lawsuits related to defects or injuries resulting from irradiation. As more and more consumers begin to understand that foodborne diseases are preventable by using available technology, their inclination towards legal action is likely to grow. Their desire for products that present a reduced risk will grow in a corresponding fashion.

Reduced "shrink"

Another major advantage is the significantly reduced "in-system" physical losses or "shrink" experienced whenever a product with improved shelf life is employed. Greater ambient stability means reduced losses throughout the entire distribution system, in the retail operation and in the home. Significant savings resulting from reduced "in-store" losses with strawberries, onions and mushrooms were reported by Corrigan [273]. Increased shelf life results in greater flexibility throughout the distribution and handling system, which generally results in measurable advantages to both the supplier and the consumer.

Greater international trade

The recently concluded GATT (General Agreement on Tariffs and Trade) agreement included a sub-agreement on the Application of Sanitary and Phytosanitary Measures. This stated that no country ". . . should be prevented from adopting or enforcing measures necessary to protect human, animal or plant life or health. . . ." It further goes on to state that the application of these measures must be based only upon sound scientific principles. This will set the stage for the much wider sales and distribution of irradiated foods, since it will be virtually impossible for any country to legitimately reject irradiated foods on sound scientific principles. Parochial politics and misinformation are thus expected to have far less of an influence on the export potential of irradiated foods than in the past. This will allow a much greater international movement of food products such as meat, seafood, poultry, fruits and vegetables.

Reduced government inspection

The recent outbreaks of foodborne disease received unprecedented national coverage in all the media. One result was the government action to institute more frequent and comprehensive inspection procedures. Although this type of action is strongly opposed by industry [274], it is certainly not surprising. Raw foods contain pathogens. With the unlikely exception of raising animals in a sterile laboratory, no level of hygienic quality control will *guarantee* freedom from pathogens in raw foods unless a step to kill them is incorporated into the process. The contemporary concept of Hazard Analysis of Critical Control Points (HAACP) had its parallel in the procedures established for the production of certified milk almost a century ago. As with certified milk, it is clear that the most meaningful procedure to quantitatively reduce the risk of pathogens is to destroy them.

The government, itself under public pressure to reduce the incidents of foodborne disease, will be reluctant to diminish its inspection function unless it is confident that adequate procedures are in place to deal with the problem. Irradiation thus has the desirable effect of reducing unwanted and government interference in the operations of manufacturers and processors. It is a case where a costly, antagonistic relationship can be remedied simply by employing a technology that everyone agrees is safe and effective.

New products

A final advantage is the potential food irradiation holds for the development of radically new products through the application of dose levels above and beyond those that were traditionally contemplated.

Although there are no toxicological concerns with foods irradiated up to a level of 10 kGy, this does not imply that such problems do exist at higher treatment levels. When the FAO/IAEA/WHO Joint Expert Committee on Irradiated Foods issued the statement that foods irradiated up to an overall average dose of 10 kGy presented no toxicological hazard, it merely reflected a range of dosage that was arbitrarily chosen for the evaluation. Studies describing irradiation treatment above 10 kGy were not included in the analysis simply because it was felt that 99% of the potential applications of food irradiation would fall in the range of treatment below that level. Indeed, sprout inhibition, ripening deferral, insect disinfestation and the reduction of spoilage microorganisms and non-spore-forming pathogens to near undetectable levels can all be achieved at doses below 10 kGy. However, this range of treatment is not sufficient to sterilize foods, and work has started on the evaluation of all research carried out up to a level of approximately 70 kGy – a dose well above that required for the full sterilization of foods [275].

It does not take much imagination to envision the range of applications that could result from a cold process to sterilize solid foods. Indeed, cold sterilization of solids may open the way to an entirely new class of shelf-stable foods. Of course, the idea is not new. A major part of the original work on irradiated military rations was directed towards a sterile end-product. The work ceased, not because of any technical barriers per se but, rather, because of the growing misperception of consumer resistance to irradiated foods. Now that the safety and value of irradiated foods are generally understood, it may be time to consider this technology more seriously. Initial work in this area has already started [276].

What would the commercial advantages of such a technology be? In the first instance, the overall cost of processing, packaging, storage, distribution and retailing would be reduced. Even home storage would be less costly. With the future limitations on the production of ozone-depleting refrigerants, it is expected that the entire refrigeration/freezing sector will encounter higher costs. Having products that could be stored at ambient temperature would therefore be a definite advantage. This advantage could also be enjoyed by foodservice operations as well as consumers.

Secondly, expansion of the concept of sterilizing foods without denaturing them will lead to a variety of new products with unparalleled hygienic quality. The concern for eating meats, fish or seafoods rare or even raw for those consumers who prefer them this way would be virtually eliminated. As a corollary, the foodservice workplace will also benefit hygienically since the raw, fresh products would no longer harbor pathogens.

It has long been recommended that immuno-compromised people, astronauts, pregnant women, children and the elderly should have access to and benefit from a more hygienic food supply. If the technology to provide a higher quality food supply was generally available, why should the rest of society not benefit from it as well?

There is no doubt that the technology for cold sterilization requires additional research to determine optimum processing conditions for each product. It has already been determined that oxidation and off-flavors can be reduced or eliminated when products are irradiated under anoxic conditions. Likewise, improved results can be obtained when products are irradiated cold or even frozen. As with any other process, experience gained through research or even simple trial and error will ultimately result in considerable product improvements.

There is a possibility that the very idea of a sterile, shelf-stable product is, in itself, a disadvantage. A sterile product that has all the organoleptic properties of the untreated product may raise the suspicions of both consumers and especially food advocates. Our age-old experience with food preservation almost precludes the existence of a sterile product that is essentially unchanged from the original. Our natural distrust of something too good to be true may prove to be the biggest stumbling block.

Another constraint may be the capital that has been invested in conventional processing technologies. Indeed, the huge investment in conventional canning was one of the main reasons for the failure of the retort pouch in the mid-1970s. Since irradiation sterilization may be a very competitive process to those methods currently in place, there may likely be a great resistance towards it on the part of those industries with heavy investments committed to conventional technology, storage and distribution systems.

Future Applications

The first line of products to consider for cold sterilization would be high value fresh foods, which are particularly subject to degradation through spoilage or pathogenic microorganisms. These would include meat, poultry, fish and seafood products, all of which now benefit from extended storage life through freezing. A primary constraint to the production of shelf-stable raw products in this category is the residual enzyme activity that results in unacceptable changes in eating quality, particularly texture. These enzymes are not destroyed by irradiation, even when sterilizing doses are employed. While these enzymes are also responsible for unacceptable changes in frozen products, the low temperatures at which they are stored results in significantly reduced activity. Work on various cuts of meat and poultry indicate such losses in texture result after ambient storage of nine months [276].

The enzymes responsible for the breakdown of meat and poultry texture are deactivated at temperatures varying from 60−72°C, the same general range employed to cook foods [277]. Thus, it would be quite difficult to inactivate the enzymes in these products without noticeable changes in their character. (The thickness and uneven geometry of many raw products in this category imply poor heat penetration, thus compounding the problem of uniform enzyme inactivation.)

Products suitable for quick heat penetration such as fillets or thin slices could be subjected to the thermal equivalent of quick blanching to destroy enzymatic activity, however, their appearance will be quite different from raw products. This does not necessarily mean that they cannot be successfully marketed. Up until they are fully defrosted, frozen products such as meats or vegetables have no resemblance to their raw counterparts, yet they are fully accepted as functionally equivalent, except that their preparation procedures are somewhat different. In the case of frozen meat and fish, preparation time is lengthened due to the requirement of a prior defrost. In the case of frozen peas or string beans, preparation time is often reduced as a result of their prior blanching.

Therefore, the possibility exists to prepare shelf-stable cuts of meat or poultry that have received a prior thermal treatment to destroy enzymatic activity. It may also be possible to employ high-pressure treatment to destroy enzyme activity prior to or even after irradiation sterilization [278,279]. Although high-pressure treatment can itself be employed as a sterilization method, product quality and process economics will ultimately dictate which technology or combination of technologies will be best for any particular product. The point is that we now appear to be at the dawn of a new era of raw or quasi-raw, shelf-stable products that would allow the consumer full flexibility to prepare foods as desired.

The problem of enzymatic degradation does not enter into precooked foods such as entrées and full dinners. These have been shown to be very adaptable to irradiation. The high quality of astronaut meals has been acknowledged for years. Recent studies have confirmed the general acceptability of such precooked foods [280]. The market for single portion, precooked foods will follow demographic trends and is expected to grow dramatically as the post-World War II generation enters its retirement period.

Beef stew, chicken Kiev, Indian curries, stuffed turkey, roast veal, poached salmon, etc., etc., all qualify for this type of processing. Once prepared and suitably packaged, they can be be cold-sterilized by irradiation. This would allow a functional shelf life which that be almost unlimited. (All that notwithstanding, it will be the vision and competence of the product formulation team that will ultimately mean the difference between foods that taste homemade or taste like they were prepared in a hospital cafeteria.)

Undoubtedly, packaging will play a critical role in this process, but even here one has to have some imagination in order to project some of the possibilities. Environmental issues, particularly recycling, will likely play a greater role in our coming packaging concepts. No packaging is more useful in this regard than those that are returned and reused after cleaning and sanitation. An obvious disadvantage of precooked meals prepared in the home compared to going to a restaurant is the initial presentation of food and the clean-up afterwards. High-quality, deposit-return packaging, including presentable, ceramic-like serving dishes, may shift movement towards more in-home consumption.

By definition, canned products are cooked. Canned soups, baked beans, fruit and vegetables, condiments, etc., will always be expected to be marketed in a precooked form. However, the quality of many canned products is markedly improved if methods to reduce the slow heat penetration to the can's geometric center are employed. Thus, processes such as ohmic heating followed by aseptic packaging are expected to enjoy growing popularity over the next number of years

[281,282]. Likewise, cold sterilization with irradiation can be expected to have excellent potential for such applications.

Irradiation will allow similar items to be packaged and sterilized without the risk of thermal over-processing. Products such as soups, fruit, vegetables and meat stews can all be cooked to the peak of organoleptic perfection, packaged, sealed and then sterilized cold. Products that are currently processed aseptically such as ultra-high temperature (UHT) juices, custard, yoghurt or even milk and cream can also be processed with minimal heat treatment. No doubt, significant product development work will be required to ensure optimum product and processing characteristics, but the potential for this type of technology is extraordinary.

A key factor involved in the feasibility evaluation of cold sterilization inevitably relates to the cost of the process. Although costs are available for the treatment of products up to the 10 kGy level, there are very limited figures currently available for sterilizing doses. Initial opinions indicate that irradiation would compete very favorably with current methodologies [283]. The cost of empty steel cans has risen dramatically in the last twenty years, as has the cost of their storage. The heavy waxed laminates that have become standard for aseptically packaged liquid products are likewise expensive and bulky. Irradiated products can be packaged in plastic pouches, laminates, plasticized cardboard or other materials, depending on a combination of cost-effectiveness and convenience. The irradiation process itself may be cheaper than the thermal treatments employed in canning or aseptic processing. There is thus an excellent possibility that irradiation can be employed in future as an alternative to canning or aseptic processing for sterile products on the basis of cost alone.

The Montreal Protocol requires that all ozone-depleting substances cease being produced by the year 2000 [284]. Although alternative refrigerants are currently moving into the market and others are planned, they are expected to increase the cost of refrigeration. As a consequence, any process that can result in a reduced need for cooling or freezing anywhere in the food chain will lessen the final product cost [285]. In those cases where refrigeration plays a significant role in the overall cost of bringing a product to the consumer, ambient shelf stability will be a decisive factor.

The functional advantage of ionizing radiation is that it can evenly penetrate a food or beverage product. The inherent effectiveness of this method makes it necessary to devise conditions of treatment that minimize all possible nutrient losses. Processing in oxygen-reduced environments and low temperatures have been effective in the past. Additional processing conditions will have to be developed for the more rigorous effects of cold sterilization. There is little doubt that, if and when products sterilized by ionizing irradiation are eventually approved for consumption, they will have to withstand the full scrutiny of government on issues of safety and nutritional quality.

In summary, all the available information suggests that it is time for the food industry to seriously consider food irradiation. Current consumer acceptability and the trend towards a growing preference for foods that will provide greater safety from foodborne diseases imply that irradiation is an ''opportunity-rich'' technology ready to be taken advantage of. Aside from its ability to deal with persistent industry problems, it will serve to increase consumer confidence in many foods and allow for much greater levels of their international trade. Irradiation will also lead to a whole range of new food products with previously unknown levels of consumer convenience and utility.

CHAPTER 9

Some Final Thoughts

A World Free of Risk

We simply do not live in a world free of risks, but risks are seldom considered in isolation. Generally, they are assessed together with any possible benefits. There is a risk to driving and to walking. There are risks in swimming and all other sports. With eating and drinking there are also a great variety of risks. Water contains chlorine and fluoride. Both are toxins. Beer, wine and alcoholic beverages all contain measurable toxins. Cassava contains cyanide, and legumes such as peas and beans contain a wide range of antinutritional toxins such as hemagglutinins, phytates and enzyme inhibitors. Potatoes contain a potent glycoalkaloid toxin called solanine, and ackee fruit contains a toxin called hypoglycin. The list of natural toxicants in foods can easily be the subject for a separate book and, in fact, a number have been written on the subject.

Transportation, work, recreation and eating are all routine and necessary activities of life which present unavoidable risks. Since, for all practical purposes they are unavoidable, our greatest defense against them is knowledge, not ignorance. A large part of humankind's activity has been devoted to increasing the benefits and decreasing the risks of all our endeavors. The risks associated with foods must be handled the same way.

What is the risk of chlorinated water compared to contaminated water? There is little opportunity of access to hand-pumped, crystal-clear water for people living in tenement houses in Brooklyn or even elegant townhouses in Chelsea. The choice is either chlorinated or contaminated water, period. Most municipalities opt for chlorinated water, not because they like the taste and certainly not because they like the chlorine producers. They choose chlorinated water because it provides a lower health risk than the alternative of contaminated water [168]. If it were ever to be discovered that chlorine treatment of water might possibly shorten our lives by a few years, the correct decision will still have been made, because even today contaminated water kills people in the millions, without waiting until they reach a ripe old age. It is simply a matter of selecting the best overall alternative based on the objective evidence available.

One of the most difficult concepts for consumers to accept is degrees of risk. The determination of food safety requires proof of a highly likely reality that the food in question will not cause consumers harm. But it cannot provide a guarantee that the food will not cause harm under any circumstances. No such guarantees exist for any food or beverage, including water. There are therefore various degrees of risk associated with foods, and professionals are employed to determine them as accurately as possible.

Most consumers find this to be a rather complicated idea, and it is. We would all rather say that something is either safe or unsafe—an absolute assessment. Anything that is unsafe should not be on the market. Unfortunately, it just isn't that simple. There are degrees of risk associated with the consumption of all foods. Determining those risks in a quantitative manner is not an easy task, and is further complicated by the great differences in the points of view between consumers and scientists.

Consumers view the determination of safety to be a simple process, while scientists know it is complicated. Consumers think personally about the issue of safety—how it affects them individually. Scientists responsible for such assessments must consider the population in its entirety in making their judgments. Scientists accept that there are uncertainties and degrees of risk, while consumers understandably look for absolute guarantees, even if they are impossible to obtain.

It is little wonder that consumers tend to look elsewhere for assurances. The problem is that those assurances are seldom built upon comprehensive knowledge, but rather on built-in biases often fostered by quacks. One of our greatest concerns is the increasing complexity of modern life and what the future might hold. Rather than face this uncertainty, there are many who council us to look to the past, not to the milestones of human development, but to a nostalgic vision of health, nature and tranquility. It is really the stuff of Hollywood and there is always an eager audience ready to accept this simplistic view. The vision may exist, but the reality never did and never will. Life in the past was not easier, it was harder. People were not healthier, they were sicker. That, however, does not stop the continual criticism of modern technologies, and the summons to return to bygone days.

Mankind is accused of tampering with nature, and thereby increasing risk to life. The argument most often brought out in recent decades is the plight of the unfortunate victims of the thalidomide tragedy. We are often reminded of it, because there are a great many advocates who wish to shake the public's confidence in the whole concept of technology. These advocates remind us of the fallibility of technology at every possible opportunity. A more condescending attitude is difficult to find. Certainly these people cannot believe that all scientists believe that science is flawless and that its record is totally unblemished. Judgments are made because decisions are needed—and they should be based on the best possible scientific evidence.

Science is a human endeavor and therefore scientific judgments can indeed be fallible. The case of thalidomide is an example of this. But progress is the result of learning from mistakes. Although it may not provide any solace, we also learn from the unfortunate victims of our mistakes. The thalidomide tragedy was the driving force behind the new science of teratology, the study of the unborn. Thalidomide was produced and sold in good faith using the most modern tests available at the time to prove its safety and usefulness. The problem was that science had not yet advanced to a state where all the intricacies could be foreseen. It never will. But does this mean that pharmaceuticals or pharmaceutical research should be abandoned? How many millions more would die as a result and who would bear the responsibility of doing nothing, or worse, turning backward? The world would still be teeming with polio victims if scientific research had not found a vaccine. We do not address the very real problems to be faced by our world by indulging in a paranoid fear of technology.

Centuries ago, the wealthy employed tasters to see if their foods were safe to eat. A great many of these poor souls didn't last long on the job. Those who could not afford the cost of a taster took their chances. Today, we leave the job to professional scientists to provide us with all the

FIGURE 9.1 The good old days.

information and controls that objective research can provide. It is only in this way that we can minimize the risks and increase the benefits of the foods we eat.

Aside from pure ignorance in eating habits, foodborne diseases are the greatest food-related risk we face. Our ability to overcome their symptoms is our only defense. Foodborne diseases are probably one of the greatest causes of deaths or diminished lives for immunocompromised and AIDS victims. This means that, in this day and age, we all do not have a fair and equal chance to avoid the misery, discomfort and even death these diseases cause, because we all have a different ability to ward off disease. Some people think this is fine, and a real test of survival of the fittest, until they themselves are struck down. Then they tend to get religion.

Foodborne diseases are serious. They must be addressed as all other diseases are. They must be eliminated wherever possible. These diseases will not be diminished by avoidance of the issue, and they will not be eliminated by uninformed charlatans no matter how eloquent or convincing they may sound. Pathogens are stupid. They don't respond to impractical speculations or moral philosophies. Pathogens simply want to survive. To do so they must infect. If we want to get rid of pathogens we have to kill them, before they kill us.

A Light at the End of the Tunnel

In the United States, there finally appears to be a light at the end of the tunnel. The first commercial facility expressly designed for the purpose of irradiating foods opened its doors in January 1992. The Vindicator Inc. plant, located in Mulberry, Florida, opened in the midst of widespread television and newspaper coverage, despite threats from Food and Water, Inc. [169–177]. Food and Water warned that it would disrupt initial shipments with large car convoys and hundreds of pickets. As it happened, one paid picket turned up at Laurenzo's, the first retailer (again) to sell the strawberries.

The strawberries, which were processed by a combination of modified atmosphere storage and irradiation went on sale January 25, 1992, in front of the parking lot at Laurenzo's Market. All boxes were labeled with a sticker showing the Radura symbol and the words "Treated by Irradiation." Despite the higher price tag of the irradiated strawberries, they outsold the non-irradiated strawberries by approximately 30% [178].

Food and Water's efforts to discourage consumers from purchasing irradiated strawberries included the charade of parading around in gas masks, presumably to ward off the effects of irradiation. In another instance, a car pulled up at a sales display and a Food and Water representative jumped out with a photographer to take a quick snapshot of the advocate holding up an anti-irradiation sign by the fruit display. Both then jumped back in the car and rode off in possession of solid proof of the hard-fought battle to discourage the sale of "nuked" strawberries.

One of the most telling experiences occurred when Vindicator fruits (strawberries and citrus) were marketed at Carrot Top Inc., a supermarket located in Northbrook, Illinois, on Chicago's North Shore. Owner-manager James Corrigan is a rather exceptional individual. He has a close relationship with his customers, and constantly strives to keep them informed of the latest in new food products. Shortly after he heard of food irradiation, he surveyed his customers through the store's newsletter to gauge their interest in irradiated foods. He received a very high rate of response with almost all people indicating that they wanted more information on the subject.

Two years later, when the first irradiated fruits became available, Carrot Top was the first retailer in the Midwest to put them on sale. Predictably, fully labeled irradiated fruits outsold the non-irradiated ones. Surprisingly, during the buy-one-get-one-free sale, most customers chose both pints to be irradiated strawberries [179].

The success of the commercial markets, combined with the wide coverage in the media, has made the head of Vindicator, Sam Whitney, something of a celebrity. Despite the commercial interests tied up in Vindicator, he is a man truly dedicated to the use of food irradiation to prevent foodborne illness. He has tirelessly toured the U.S., often facing abuse from anti-irradiation advocates in order to get his message across [180].

Why is it that fairly small companies such as Vindicator (recently renamed to Food Technology Service, Inc.), Laurenzo's and Carrot Top have taken the lead in supplying irradiated foods to consumers rather than the much larger national giants? When examined more closely it appears that they have a number of things in common, even though they are rather different operations. In the first instance all three are fairly small companies compared to the majors in their business. Another characteristic is that they are operated by private entrepreneurs who are very close to their clients and deal directly with them. These companies do not have large public relations departments to separate the chief executive from the clients. This proximity to customers provides these managers with a degree of understanding and confidence in dealing with clients that is often lacking in larger companies. It is unfortunate that the trailblazing role is always taken by these small companies, when the larger ones have much greater resources to apply to the problem. Hopefully, history will credit the pioneering efforts of these entrepreneurs in paving the way forward to significantly reduce the frequency of foodborne disease.

Thus far, every single market trial has clearly demonstrated a consumer preference for irradiated foods. The evidence even indicates that consumers are willing to pay a premium for irradiated foods once they understand their advantages in terms of hygiene, shelf life and eating quality [181]. Normally, this fact alone would be sufficient incentive for food company executives to press for irradiated foods.

It therefore seems extraordinary that New Wave advocates, who have resorted to the theatrics of donning gas masks, appear to have influenced the opinions of chief executives more than the experts trained to accurately gauge consumer response. All the signs appear to indicate that they are not only denying their customers a choice of products, but they are also letting down their shareholders by refusing to market a product of proven preference.

Even the single highest public health authority in the United States has given unqualified support for food irradiation. Dr. James O. Mason, the Head of the Public Health Service and Assistant Secretary for Health, has gone on record saying that food irradiation has for too long languished in disuse—"an ironic situation for a nation that prides itself on scientific know-how and a 'can-do' spirit" [182].

On September 21, 1992, official U.S. approval for the irradiation of poultry was enacted. The countless studies and reports that were witness to the safety and efficacy of food irradiation finally resulted in a decision which will have a profound effect on the integrity of our food supply. Sure enough, the two retail supermarkets which led the move to bring irradiated poultry to consumers—poultry that for the first time provided consumers with a quantum leap in food safety—were Laurenzo's and Carrot Top Inc. [267]. It remains to be seen if, and for how long, the giant poultry producers and supermarkets will delay bringing safer irradiated poultry to the consumer.

There is little doubt, however, that the imminent approval of shellfish will result in an immediate commercial marketing program. Shellfish producers hope to have an immediate expansion of sales with irradiated products. Even NASA has filed a petition for the irradiation of beefsteaks to be consumed by all the astronauts in future [183]. More recently (June 1994), a comprehensive petition for the irradiation treatment of meat was submitted to the FDA. The key areas of focus were the application of a maximum dose of 4.5 kGy for fresh intact and ground meat, as well as maximum doses of 70 kGy for the same materials in frozen form [286].

Thus the entire issue has finally moved from the laboratory and pilot plant to the commercial processor and retailer. It will ultimately be in their court to provide clients with the options.

A very recent tragic incident became the pivotal event that finally forced government and industry to address the issue of foodborne disease more aggressively. In mid-January of 1993, approximately 540 consumers in the states of Washington, Idaho, California and Nevada were stricken by a severe outbreak of *E. coli* food poisoning. The particular microorganism responsible was *E. coli* O157:H7, commonly referred to as ECO157. As mentioned earlier, this is a particularly virulent form of *E. coli* and, in the latest food poisoning-related incident, was found to be responsible for a large number of critical casualties, in addition to the deaths of four young children [184]. Victims not much older than infants were forced to undergo prolonged dialysis due to hemolytic uremia and some even suffered strokes. One of the fatalities, a two-year-old child, died of heart failure. The Center for Disease Control (CDC) in Atlanta, Georgia, carried out a follow-up investigation to determine the proportion of young victims that developed chronic renal failure in order to identify those avoidable risks that could make kidney failure a likely prospect. (In fact, the CDC should be highly lauded for its very rapid and professional response to this incident. If not for their actions in quickly identifying those day care centers with ECO157-positive children, there would probably have been many more victims resulting from person-to-person spread.)

The food poisoning outbreak was linked to the consumption of hamburgers at the Jack-in-the-Box restaurants in the affected regions. This tragic incident revealed a series of weaknesses in the food system as well as the type of dynamics that come into play when such a misfortune occurs.

Like most similar large chains, the Jack-in-the-Box restaurants contract out the production of their raw hamburgers to centralized meat processors. In this particular incident, The Vons Companies was the manufacturer of the implicated frozen patties. It is quite likely that the original source of ECO157-contaminated meat was not Vons, but could have been a packing house or a particular herd or farm. The fact is that all these locations might serve as potential sources of various pathogens, including ECO157. The mere presence of ECO157 in the frozen hamburgers, however, was not the entire cause of this outbreak. As it turned out, the hamburger cooking procedures at Jack-in-the-Box were not sufficient to destroy the ECO157 present in the incoming patties. Thus it was a combination of factors which led to this tragic event and, given the technology and legislation currently in place, it is extremely difficult to point the finger of technical culpability in any one direction.

Meat and poultry inspection legislation did not consider products containing bacteria (including pathogens) to be adulterated. This is simply because bacteria occur naturally in these products. Since

ECO157 does not cause illness in the host animals (much in the same way that most *Salmonella* do not make poultry ill), the presence of these organisms in the food supply is not at all surprising. Animals appear healthy and normal in all respects and easily pass visual inspection. In fact, ECO157 is so virulent to humans that the small number of bacteria required to cause a food poisoning incident may easily be missed in routine meat quality control tests. At the time of the incident, the Food and Drug Administration recommendation for minimum cooking temperature at retail food service outlets was 140°F. This temperature is insufficient to kill all the pathogens present. Since the outbreak, however, most hamburger chains have raised their minimum cooking temperature to 155°F.

As could be expected, fingers were pointed everywhere in an attempt to assess guilt. Whose fault could it have been, when all involved appeared to have complied with the law? Perhaps the fault lay in the law itself. Bacteria (including ECO157) occur just as naturally in milk as they do in poultry and beef. Yet, in the case of milk, we do not hide behind this natural occurrence to excuse the presence of pathogens. We pasteurize milk to ensure that any bacteria originally present cannot cause illness when the final product is consumed.

Almost without exception, all those involved in the public debate which followed the Jack-in-the-Box incident stated that the only answer to this problem is proper cooking. In other words, the final handler (be it the cook in the restaurant or at home) bears the final responsibility for the safety of the consumer. While no one can deny the need to properly cook foods, it does seem bizarre that in a modern country where both the available technology and infrastructure are capable of providing the consumer a greater degree of safety, the final arbiter is the cook! Using this same philosophy, we might as well go back to boiling milk or drinking water. The sophisticated technology and infrastructure we have are supposed to eliminate the need for leaving consumers to their own devices in order to guarantee their safety.

The gravity of the hamburger poisoning incident resulted in meetings of the U.S. Senate Subcommittee on Agricultural Research, Conservation, Forestry and General Legislation. Both the Secretary of Agriculture and the Administrator of the Food Safety and Inspection Service (USDA) were among those questioned. One of the measures to be instituted was an immediate priority given for research to support a petition for FDA approval of irradiation for fresh ground beef and beef trimmings. In the dozens of newspaper reports that followed, the Secretary of Agriculture indicated that legislation allowing the irradiation of beef would be considered. A program to increase microbiological testing in order to establish reliable baseline figures for bacteria levels in beef, poultry and pork was to be instituted along with a raft of other measures.

Perhaps the most significant action was the National Cattlemen's Association's (NCA) endorsement of a policy supporting irradiation research. The policy stated that "Upon the determination of positive results, the NCA will encourage USDA's Food Safety and Inspection Service to establish irradiation procedures for beef." Rather than sitting back and saying there is no problem, the NCA wanted to aggressively move forward to ensure the elimination of future problems. The most recent results clearly demonstrate the effectiveness of γ-irradiation in the elimination of *E. coli* O157:H7 in meats [234].

In December 1993, the American Medical Association approved a statement endorsing irradiation as part of a "comprehensive food safety program based on good manufacturing practices [287]." As pointed out in Chapter 8, "Irradiation and the Food Industry," the American Gastroenterological Association Foundation held a National Consensus Conference in Washington in July 1994, to address the problem of *E. coli* O157:H7. The conclusions, markedly in favor of wide-scale irradiation, were quoted by major newspapers around the country [288,289]. One newspaper in Milwaukee was able to combine the Consensus Conference story with an actual hamburger poisoning outbreak that had just occurred in the city [290].

The appearance of stories on food poisoning outbreaks has become a commonplace occurrence [291,292], and food irradiation has clearly become a viable option to deal with this problem. A recent article in the Personal Health section of the *New York Times* is entitled "Fears of Food Irradiation Are Greatly Exaggerated [293]." Another editorial in the respected newspaper, *The Oregonian*, clearly states that food irradiation holds the key to healthier meat, seafood and poultry [296]. It thus appears that, since the first edition of *Food Irradiation* was published, a corner has been turned in the progress of this technology. As with pasteurization, it will be public opinion that will drive the demand for better and safer foods.

However, there still remains some caution concerning potential consumer resistance to irradiated foods in the marketplace. This caution is not well-grounded since all market tests have demonstrated a preference rather than a resistance for products such as irradiated fruits and vegetables. When consumers are presented fully labeled, irradiated poultry, beef or pork and they understand that these products pose a greatly decreased risk from pathogenic organisms, there is no doubt as to their preferences. Restaurant chains will doubtlessly feel the same way.

The fact that so many influential people still feel that consumers are against irradiated foods, despite all the published evidence to the contrary, is itself rather interesting. It is clear that those who have spread myth-information about irradiation have indeed played a vital role in delaying the decisions that would result in consumer access to safe, irradiated foods. In a recent letter to the *Financial Times*, Ms. Maria Elena Hurtado, the Global Policy and Campaigns Director of the International Organization of Consumer Unions (IOCU) stated that, "Irradiating them will leave grain full of dead insects or mice" [294]. A large supermarket chain recently replied to a letter of mine stating that they "will not sell any irradiated produce, meat, seafood. We listen to our customers and our customers are telling us that they do not want these irradiated products" [295]. The poor consumer is thus consigned to the fate of the bumblebee, which, according to wing-loading principles and the laws of aerodynamics, cannot fly. Fortunately, the bumblebee is not aware of this and flies anyway. Despite the negative statements and attitudes on the part of those purportedly working in the interests of the consumers, it is truly a wonder that paying consumers have made irradiated foods such an outstanding success wherever they have been introduced.

Although it is useless to speculate as to whether the ECO157 hamburger tragedy could have been avoided if beef irradiation had been in place, it is clear that no one will stand up to take the responsibility for the delay. No one who influentially resisted and delayed pasteurization stood to accept the responsibility for the thousands who died needlessly until it was finally instituted. It will be the same for irradiated foods.

Thus we have come full circle. A century after it was first employed, food irradiation is finally poised for full-scale implementation. Ironically, it will shortly be the centenary of Pasteur's death (1995). Yet, as happens so often, it has taken a preventable, needless tragedy to set actions in motion.

A Free and Informed Choice

The scientific community has declared that food irradiation is a safe, practical and convenient method for preserving a wide variety of foods and decreasing their risk of spreading disease. The technology has been tested for decades, and found to be unique in its ability to accomplish certain tasks, particularly the reduction of pathogens and spoilage microorganisms. At this point in time, its use in the prevention of foodborne diseases, its ability to control sprouting, and its functional application in quarantine treatment and prevention of postharvest losses is unique and, in many cases, far more practical than other methods.

Yet, there is no doubt that its name, its novelty, its ability to provide consumers with better products than before, and most of all, its association with the nuclear issue make some people suspicious of it. This is not a unique phenomenon since a similar resistance was encountered with pasteurization, and with microwave ovens. There will always be consumers who prefer more traditional ways of doing things because they find security in tradition and in the past. This is why cast iron utensils and wood burning stoves can still be found in some stores. It is why people sometimes go to rural areas to buy "farm-fresh" milk. It is also why there are many households that find no need for a microwave oven. But that is good, because it reflects our individual views. It also reflects our right of choice. And this has been the central issue of this book.

People must be allowed free choices, particularly when the weight of evidence indicates a benefit for a particular product or technology. And there is little doubt that the reduction of foodborne disease is an immense benefit to us. So is the reduction of food waste. Elevating free choices to free and informed choices is dependent upon the efforts made by consumers and all those who serve them to ensure the provision of solid, unbiased information. It must also be accepted that the mere provision of a free and informed choice will never be a guarantee that the most rational decisions will be made by consumers. It is difficult to convince some people to refrain from smoking, or to stop wealthy Oriental businessmen from poisoning themselves with toxic fish, or to discourage people from going back to nature and drinking raw milk or eating questionable mushrooms. Free choice and information are not implicit guarantees of safety. But they can allow those consumers who wish to, the possibility of decreasing the routine risks in life to a level which they find more acceptable.

To quote Jim Corrigan of Carrot Top Inc., "It's not really up to me as a retailer to say you have to eat this or you should eat that. Rather, it's for me to offer my customers the option, and keep getting them the information they need to make up their own minds. I certainly believe

my customers are intelligent enough to make that kind of decision''
[185].

Unfortunately, Jim Corrigan is the exception rather than the rule and
there is currently a very limited ability for consumers, who so choose,
to obtain the degree of safety and convenience in food choices which
modern technology can provide.

Consumers have been poorly treated by the very people and institutions set up to serve them. Politics, self-interest, and ignorance are the
time-honored means that have combined to long deny the consumer fair
treatment in their food choices.

It's time to do something about it, don't you think?

REFERENCES

1 Vallery-Radot, R. (No date indicated.) *The Life of Pasteur*. New York: Garden City Publishing, p. ix.

2 Flat Earth Research Society International, Box 2533, Lancaster, CA 93539, USA.

3 Considine, D. M., ed. 1976. *Van Nostrand's Scientific Encyclopedia, 5th Edition*. New York: Van Nostrand Reinhold Co.

4 Thatcher, V. F., ed. 1971. *The New Webster Encyclopedic Dictionary of the English Language*. Chicago: Consolidated Book Publishers.

5 Weckel, K. G. and H. C. Jackson. 1939. "The Irradiation of Milk," *Wisconsin Agricultural Experiment Station (Madison) Research Bulletin* (136):1−55.

6 Golden, M. N. H. and D. Ramdath. 1986. "Free Radicals in the Pathogenesis of Kwashiorkor," in *Proceedings of the XII International Congress of Nutrition*, T. G. Taylor and N. K. Jenkins, eds., London: John Libby and Company, pp. 597−598.

7 Urbain, W. M. 1986. *Food Irradiation*. Orlando: Academic Press, Inc.

8 Josephson, E. S. 1983. "An Historical Review of Food Irradiation," *Journal of Food Safety*, 5(4):161.

9 Diehl, J. F. 1990. *Safety of Irradiated Foods*. New York: Marcel Dekker, Inc.

10 Minsch, F. 1896. *Münch Med Wochensch*, 5:101; 9:202.

11 Lieber, H. U.S. Patent 788480, April 25, 1905. Appleby, J. and A. J. Banks, Brit. Patent No. 1609, Jan. 26, 1905.

12 Runner, G. A. 1916. "Effect of Roentgen Rays on the Tobacco or Cigarette Beetle and the Results with a New Form of Roentgen Tube," *Journal of Agricultural Research*, 6:383.

13 Schwartz, B. 1921. "Effect of X-Rays on Trichinae," *Journal of Agricultural Research*, 20:845.

14 "WHO, Wholesomeness of Irradiated Foods," Technical Report Series 659, Geneva, 1981.

15 Sommer, H. H. 1952. *Market Milk and Related Products, 3rd Edition*. Madison, Wisconsin: Published by Author.

16 Pasteur, L. 1866. *Études Sur Le Vin*, L'Imprimerie Impériale, Paris.

17 Kilbourne, C. H. 1916. *The Pasteurization of Milk*. New York: John Wiley and Sons.

18 Rosenau, M. J. 1913. *The Milk Question*. Boston: Houghton Mifflin Company.

19 Hall, C. W. and G. M. Trout. 1968. *Milk Pasteurization*. Westport, Connecticut: The Avi Publishing Company.

20 Wiley, H. W. 1917. *Foods and Their Adulteration, 3rd Edition*. Philadelphia: P. Blakiston's Son and Co.

21 Wilson, G. S. 1942. *The Pasteurization of Milk*. London: Edward Arnold and Co.

22 Humphrey, T. J. and R. J. C. Hart. 1988. "*Campylobacter* and *Salmonella* Contamination of Unpasteurized Cow's Milk on Sale to the Public," *Journal of Applied Bacteriology*, 65:463.

23 Forbes, G. I., J. C. M. Sharp, P. W. Collier, W. J. Reilly and G. M. Paterson. 1986. "Food Epidemiology—Milk Borne Salmonellosis Affecting Farming Communities in Scotland," *Environmental Health*, 94(10):269.

24 Steenbock, H. 1924. "The Induction of Growth Promoting and Calcifying Properties in a Ration by Exposure to Light," *Science*, 60:224.

25 Hess, A. F. and M. Weinstock. 1924. "Antirachitic Properties Imparted to Inert Fluids and to Green Vegetables by Ultra-Violet Radiation," *Journal of Biological Chemistry*, 62:301.

26 Searle, A. F. J. and P. McAthey. 1989. "Treatment of Milk by Gamma Irradiation—Effect of Anoxia on Lipid Peroxidation and the Survival of *Pseudomonas aeroginosa*," *Journal of the Science of Food and Agriculture*, 48:361.

27 Garthwright, W. E., D. L. Archer and J. E. Kvenberg. 1988. "Estimates of Incidence and Costs of Intestinal Infectious Disease in the United States," *Public Health Reports*, 103(2):107.

28 Schmidt, A. M. 1975. "Food and Drug Law: A 200 Year Perspective," *Nutrition Today*, 10(4):32.

29 Ministry of Agriculture. 1987. *Fisheries and Food, Survey of Consumer Attitudes to Food Additives, Vol. 1*. HMSO, London.

30 Jackson, G. J., C. F. Langford and D. L. Archer. 1991. "Control of Salmonellosis and Similar Foodborne Infections," *Food Control*, 2(1):26.

31 D'Aoust, J. Y. 1989. *Salmonella*, in *Foodborne Bacterial Pathogens*, M. P. Doyle, ed., New York: Marcel Dekker, pp. 327—445.

32 Palmer, S. R., J. E. M. Watkeys, I. Zamiri, P. G. Hutchings, C. H. L. Howells and J. F. Skone. 1990. "Outbreak of *Salmonella* Food Poisoning amongst Delegates at a Medical Conference," *Journal of the Royal College of Physicians of London*, 24(1):26.

33 Unpublished data.

34 Zottola, E. A. and L. B. Smith. 1990. "The Microbiology of Foodborne Disease Outbreaks: An Update," *Journal of Food Safety*, 11:13.

35 Bean, N. H. and P. M. Griffin. 1990. "Foodborne Disease Outbreaks in the United States, 1973–1987: Pathogens, Vehicles, and Trends," *Journal of Food Protection*, 53(9):804.

36 Todd, E. C. D. 1989. "Preliminary Estimates of Costs of Foodborne Disease in Canada and Costs to Reduce *Salmonellosis*," *Journal of Food Protection*, 52(8):586.

37 Rubery, E. D. 1990. "Role of the Department of Health in the Implementation of the New UK Food Safety Bill," *Food Control* (Oct.):295.

38 Jackson, J. J., C. F. Langford and D. L. Archer. 1991. "Control of Salmonellosis and Similar Foodborne Infections," *Food Control*. 2(1):26.

39 D'Aoust, J. Y. 1991. "Pathogenicity of Foodborne *Salmonella*," *International Journal of Food Microbiology*, 12:17.

40 Soper, G. A. 1919. *Military Surgery*, 45:1.

41 Edel, W., M. van Schothorst and E. H. Kampelmacher. 1976. "Epidemiological Studies on *Salmonella* in a Certain Area ("Walcherin Project"), 1. The Presence of *Salmonella* in Man, Pigs, Insects, Seagulls and in Foods and Effluents," *Zbl. Bakt. Hyg. 1. Abt. Orig. A* (National Institute of Health, Bilthoven, The Netherlands), 325:476.

42 Fain, A. R. 1990. *Scope* (Silliker Laboratories), 5(1):1.

43 Lillard, H. S. 1989. "Factors Affecting the Persistence of *Salmonella* during the Processing of Poultry," *Journal of Food Protection*, 52(11):829.

44 Lindsay, R. E., W. A. Krissinger and B. F. Fields. 1986. "Microwave versus Conventional Oven Cooking of Chicken: Relationship of Internal Temperature to Surface Contamination of *Salmonella typhimurium*," *Journal of the American Dietetic Association*, 86:373.

45 Schnepf, M. and W. E. Barbeau. 1989. "Survival of *Salmonella typhimurium* in Roasting Chickens Cooked in a Microwave, Convection Microwave, and a Conventional Electric Oven," *Journal of Food Safety*, 9:245.

46 Bergdoll, M. S. 1989. "*Staphylococcus aureus*," in *Foodborne Bacterial Pathogens*, M. P. Doyle, ed., New York: Marcel Dekker, pp. 463–523.

47 Tranter, H. S. 1990. "Foodborne Staphylococcal Illness," *The Lancet*, 336:1044.

48 Genigeorgis, C. A. 1989. "Present State of Knowledge on Staphylococcal Intoxication," *International Journal of Food Microbiology*, 9:327.

49 Sakaguchi, G. 1979. *Botulism, in Food-Borne Infections and Intoxications, Second Edition*, H. Reimann and F. L. Bryan, eds., New York: Academic Press.

50 Sugiyama, H. 1990. *Botulism, in Foodborne Diseases*, D. O. Cliver, ed., San Diego: Academic Press.

51 Conner, D. E., V. N. Scott, D. T. Bernard and D. A. Kautter. 1989. "Potential

Clostridium botulinum Hazards Associated with Extended Shelf-Life Refrigerated Foods: A Review," *Journal of Food Safety*, 10:131.

52 Lambert, A. D., J. P. Smith and K. L. Dodds. 1991. "Effect of Headspace CO_2 Concentration on Toxin Production by *Clostridium botulinum* in MAP Irradiated Fresh Pork," *Journal of Food Protection*, 54(8):588–592.

53 Lund, B. M. 1990. "Foodborne Disease due to *Bacillus* and *Clostridium* Species," *The Lancet*, 336:983.

54 de Guzman, A. M. S., B. Micalizzi, C. E. T. Pagano and D. Giménez. 1990. "Incidence of *C. perfringens* in Fresh Sausages in Argentina," *Journal of Food Protection*, 53:173.

55 Skirrow, M. B. 1991. "Epidemiology of *Campylobacter enteritis*," *International Journal of Food Microbiology*, 12:9.

56 Butzler, J.-P. and J. Oosterom. 1991. "*Campylobacter*: Pathogenicity and Significance in Foods," *International Journal of Food Microbiology*, 12:1.

57 Skirrow, M. B. 1990. "*Campylobacter*," *The Lancet*, 336:921.

58 Hudson, S. J., A. O. Sobo, K. Russel and N. F. Lightfoot. 1990. "Jackdaws as Potential Source of Milk-Borne *Campylobacter jejuni* Infection," *The Lancet*, 335:1160.

59 Lovett, J. 1989. "*Listeria monocytogenes*," in *Foodborne Bacterial Pathogens*, M. P. Doyle, ed., New York: Marcel Dekker, pp. 283–310.

60 Silliker, J. H. 1986. "*Listeria monocytogenes*, in Bacteria in the News," *Food Technology*, 24(August).

61 Harwig, J., P. R. Mayers, B. Brown and J. M. Farber. 1991. "*Listeria monocytogenes* in Foods," *Food Control*, 4:66.

62 Schlech, W. F., III, P. M. Lavigne, R. A. Bortolussi, A. C. Allen, E. V. Haldane, A. J. Wort, A. W. Hightower, S. E. Johnson, S. H. King, E. S. Nicholls and C. V. Broome. 1983. "Epidemic Listeriosis–Evidence for Transmission by Food," *New England Journal of Medicine*, 308:203.

63 Johnson, J. L., M. P. Doyle and R. G. Cassens. 1990. "*Listeria monocytogenes* and Other *Listeria* spp. in Meat and Meat Products–A Review," *Journal of Food Protection*, 53(1):81.

64 Breer, C. and K. Schopfer. 1989. "Listerien in Nahrungsmitteln," *Schweizerische Medizinische Wochenschrift*, 119(10):306.

65 Vanderlinde, P. B. and F. H. Grau. 1991. "Detection of *Listeria* spp. in Meat and Environmental Samples by an Enzyme-Linked Immunosorbent Assay (ELISA)," *Journal of Food Protection*, 54(3):230.

66 Hollywood, N. W., Y. Varabioff and G. E. Mitchell. 1991. "The Effect of Microwave and Conventional Cooking on the Temperature Profiles and Microbial Flora of Minced Beef," *International Journal of Food Microbiology*, 14:67.

67 Dupont, H. L. 1986. "Consumption of Raw Shellfish–Is the Risk Now Unacceptable?" *New England Journal of Medicine*, 314:707.

68 Jianxiang, W., T. Yiwei, Q. W. G. Yifang, X. Jianxin and Zhiyi. 1988. "Seroepidemiological Survey of Viral Hepatitis A during an Epidemic in Shanghai," *Acta Academiae Medicinae Shanghai*, 15:379.

69 Collee, J. G. 1990. "Bovine Spongiform Encephalitis," *The Lancet*, 336:1300.

70 Tartakow, I. J. and J. H. Vorperian. 1981. *Foodborne and Waterborne Diseases*. Westport: Avi Publishing Co.

71 Casemore, D. P. 1990. "Foodborne Protozoal Infection," *The Lancet*, 336:1427.

72 Kirk, J. A. and J. S. Remington. 1978. "Toxoplasmosis in the Adult—An Overview," *New England Medical Journal*, 298:550.

73 Holley, H. P. and C. Dover. 1986. "*Cryptosporidium*: A Common Cause of Parasitic Diarrhoea in Otherwise Healthy Individuals," *Journal of Infectious Diseases*, 153:365.

74 Hoskin, J. C. and R. E. Wright. 1991. "*Cryptosporidium*: An Emerging Concern for the Food Industry," *Journal of Food Protection*, 54(1):53.

75 Archer, D. L. and J. E. Kvenberg. 1985. "Incidence and Cost of Foodborne Diarrhoeal Disease in the United States," *Journal of Food Protection*, 48(10):887.

76 Todd, E. C. D. 1989. "Preliminary Estimates of Costs of Foodborne Disease in the United States," *Journal of Food Protection*, 52(8):595.

77 Todd, E. C. D. 1989. "Preliminary Costs of Foodborne Disease in Canada and Costs to Reduce Salmonellosis," *Journal of Food Protection*, 52(8):586.

78 Hecht, A. 1991. "Preventing Food-Borne Illness," *FDA Consumer*, 18(Jan-Feb).

79 Urbain, W. M. 1983. "Radurization and Radicidation: Meat and Poultry," in *Preservation of Food by Ionizing Radiation*, E. S. Josephson and M. S. Peterson, eds., Boca Raton: CRC Press, Inc.

80 Brake, R. J., K. D. Murrell, E. E. Ray, J. D. Thomas, B. A. Muggenburg and J. S. Sivinski. 1985. "Destruction of *Trichinella spiralis* by Low Dose Irradiation of Infected Pork," *Journal of Food Safety*, 7:127.

81 Nickerson, J. T. R., J. J. Licciardello and L. J. Ronsivalli. 1983. "Radurization and Radicidation: Fish and Shellfish," in *Preservation of Food by Ionizing Radiation*, E. S. Josephson and M. S. Peterson, eds., Boca Raton: CRC Press, Inc.

82 Nerkar, D. P. and J. R. Bandekar. 1990. "Elimination of *Salmonella* from Frozen Shrimp by Gamma Radiation," *Journal of Food Safety*, 10:175.

83 Lefebvre, N., C. Thibault and R. Charbonneau. 1992. "Improvement of Shelf-Life and Wholesomeness of Ground Beef by Irradiation, 1. Microbial Aspects," *Meat Science*, 32(2):203–213.

84 Jingtian, Y., G. Xinhua, G. Guoxing and Y. Guichun. 1988. "Studies of Soy Sauce Sterilization and Its Special Flavour Improvement by Gamma-Ray Irradiation," *Radiation Physics and Chemistry*, 32:209.

85 Grant, I. R. and M. F. Patterson. 1991. "Effect of Irradiation and Modified Atmosphere Packaging on the Microbial Safety of Minced Pork Stored under Temperature Abuse Conditions," *International Journal of Food Science and Technology*, 26(5):521–533.

86 Hau, L. B., M. H. Liew and L. T. Yeh. 1992. "Preservation of Grass Prawns by Ionizing Radiation," *Journal of Food Protection*, 55(3):198–202.

87 Thibault, C. and R. Charbonneau. 1991. "Extension of Shelf-Life of Atlantic Cod (*Gadus morhua*) Fillets with the Aid of Ionizing Radiation, II. Evaluation of Odour and Chemical Indicators," *Science et Aliments*, 11(2):249–261.

88 Grohmann, G. S., P. J. Murphey, G. Christopher, G. Auty and H. B. Greenberg. 1981. "Norwalk Virus *Gastroenteritis* in Volunteers Consuming Depurated Oysters," *Australian Journal of Experimental Biology and Medical Sciences*, 59:219.

89 Mallett, J. C., L. E. Beghian, T. G. Metcalf and J. D. Kaylor. 1991. "Potential of Irradiation Technology for Improved Shellfish Sanitation," *Journal of Food Safety*, 11:231.

90 Notermans, S. and E. H. Kampelmacher. 1974. "Attachment of Some Bacterial Strains to the Skin of Broiler Chickens," *British Poultry Science*, 15:573.

91 Lillard, H. S. 1989. "Incidence and Recovery of *Salmonellae* and Other Bacteria from Commercially Processed Poultry Carcasses at Selected Pre- and Post-Evisceration Steps," *Journal of Food Protection*, 52(2):89.

92 Public Health Laboratory Service. 1990. "Memorandum of Evidence to the Agriculture Committee Inquiry on *Salmonella* in Eggs," Public Health Laboratory Service, *Microbiological Digest*, 6:1, quoted in Roberts, D. 1990. "Sources of Infection: Food," *The Lancet*, 336:859.

93 Bailey, J. S., A. C. Nelson and L. C. Blankenship. 1991. "A Comparison of an Enzyme Immunoassay, DNA Hybridization, Antibody Immobilization, and Conventional Methods for Recovery of Naturally Occurring *Salmonellae* from Processed Broiler Carcasses," *Journal of Food Protection*, 54(5):354.

94 Reilly, W. J., G. I. Forbes, J. C. M. Sharp, S. I. Oboegbulem, P. W. Collier and G. M. Paterson. 1988. "Poultry-Borne *Salmonellosis* in Scotland," *Epidemiology and Infection*, 101:115.

95 Lapidot, M., I. Klinger, E. Eisenberg, R. Padova, I. Ross and S. Saveneau. (In press.) "Application of Ionizing Radiation in Israel to Produce Safe Foods and Reduce Food-Borne Infections," presented at the Agricultural Research Institute Conference "Safeguarding the Food Supply through Irradiation Processing Techniques," Orlando, Florida, 25–31 October, 1992.

96 Thayer, D. W., S. Songprasertchai and G. Boyd. 1991. "Effects of Heat and Ionizing Radiation on *Salmonella typhimuruim* in Mechanically Deboned Chicken Meat," *Journal of Food Protection*, 54(9):718–724.

97 Roberts, T. 1985. "Microbial Pathogens in Raw Pork, Chicken, and Beef: Benefit Estimates for Control Using Irradiation," *American Journal of Agricultural Economics*, 67:957.

98 Narvaiz, P., G. Lescano, E. Kairiyama and N. Kaupert. 1989. "Decontamination of Spices by Irradiation," *Journal of Food Safety*, 10:49.

99 Hatton, T. T., R. H. Cubbege, L. A. Risse, P. W. Hale, D. H. Spalding and W. F. Reeder. 1982. "Phytotoxicity of Gamma Irradiation of Florida Grapefruit," *Proceedings of the Florida State Horticultural Society*, 95:232.

100 Lester, G. E. and D. A. Wolfenbarger. 1990. "Comparisons of Cobalt-60 Gamma Irradiation Dose Rates on Grapefruit Flavedo Tissue and on Mexican Fruit Fly Mortality," *Journal of Food Protection*, 53(4):329.

101 Heather, N. W., R. J. Corcoran and C. Banos. 1991. "Disinfestation of Mangoes with Gamma Irradiation against Two Australian Fruit Flies (*Diptera, Tephritidae*)," *Journal of Economic Entomology*, 84(4):1304–1307.

102 McLauchlan, R. L., G. E. Mitchell, G. I. Johnson, S. M. Nottingham and K. M. Hammerton. 1992. "Effects of Disinfestation-Dose Irradiation on the Physiology of Tai So Lychee," *Postharvest Biology and Technology*, 1(3):273–281.

103 Lester, G. 1989. "Gamma Irradiation, Hot Water and Imazalil Treatments on Decay Organisms and Physical Quality of Stored Netted Muskmelon Fruit," *Journal of Food Safety*, 10:21.

104 Dalziel, J. and H. J. Duncan. 1980. "Studies on Potato Sprout Suppressants. 4. The Distribution of Tecnazine in Potato Tubers and the Effect of Processing on Residue Levels in Treated Tubers," *Potato Research*, 23:405.

105 Poole, S. E., P. Wilson, G. E. Mitchell and P. A. Wills. 1990. "Storage Life of Chilled Scallops Treated with Low Dose Irradiation," *Journal of Food Protection*, 53(9):763–766.

106 Larrigaudiere, C., A. Latche, J. C. Pech and C. Triantaphylides. 1991. "Relationship between Stress Ethylene Production Induced by Gamma Irradiation and Ripening of Cherry Tomatoes," *Journal of the American Society for Horticultural Science*, 116(6):1000–1003.

107 Ayensu, E. S. 1990. "Application of Irradiation Technology and Africa's Food Needs," Report prepared for the United Nations Economic Commission for Africa, Natural Resources Division and the African Regional Centre for Technology.

108 Aworh, O. C. 1986. "Food Irradiation in Nigeria–Problems and Prospects," *Nigerian Food Journal*, 4(1):131.

109 Takyi, E. E. K. and I. K. Amuh. 1979. "Wholesomeness of Irradiated Cocoa Beans. The Effect of Irradiation on the Chemical Constituents of Cocoa Beans," *Journal of Agricultural and Food Chemistry*, 27(5):979.

110 Adesuyi, S. A. 1976. "The Use of Gamma Irradiation for Control of Sprouting in Yams (*D. rotundata*) during Storage," *Nigerian Journal of Plant Protection*, 2:34.

111 Todd, E. C. D. 1987. "Impact of Spoilage and Foodborne Diseases on National and International Economies," *International Journal of Food Microbiology*, 4:83.

112 Foster, A. 1991. "Consumer Attitudes to Irradiation," *Food Control*, 2(1):12.

113 Kahn, H. 1979. "The World at a Turning Point: New Class Attitudes," in *Critical Food Issues of the Eighties*, M. Chou and D. P. Harmon, eds., New York: Pergamon Press.

114 Taylor, J. 1988. "Consumer Views on Acceptance of Irradiated Food," Keynote address at the Joint FAO/IAEA/WHO/ITC-UNCTAD/GATT International Conference on the Acceptance, Control of and Trade in Irradiated Food, Geneva, Switzerland, 12–16 December, 1988 (available from WHO, Geneva).

115 Murray, D. R. 1990. *Biology of Food Irradiation*. Taunton, Somerset: Research Studies Press Ltd.

116 Priyadarshini, E. and P. G. Tulpule. 1976. "Aflatoxin Production on Irradiated Foods," *Food and Cosmetic Toxicology*, 14:293.

117 Priyadarshini, E. and P. G. Tulpule. 1979. "Effect of Graded Doses of Gamma-Irradiation on Aflatoxin Production by *Aspergillus parasiticus* in Wheat," *Food and Cosmetic Toxicology*, 17:505.

118 Ogbadu, G. 1980. "Influence of Gamma Irradiation on Aflatoxin B_1 Production by *Aspergillus flavus* Growing on Some Nigerian Foodstuffs," *Microbios*, 27:19.

119 Chang, H. G. and P. Markakis. 1982. "Effect of Gamma Irradiation on Aflatoxin Production in Barley," *Journal of the Science of Food and Agriculture*, 33:559.

120 Rodriguez, M. and Y. A. Rodriguez. 1983. "Reduction of Aflatoxin Formation in Peanuts by Gamma-Irradiation" (Spanish), *Cienca y Technica en la Agricultura, Veterinaria*, 5:103.

121 Behere, A. G., A. Sharma, S. R. Padwaldesai and G. B. Nadkarni. 1978. "Production of Aflatoxins during Storage of Gamma Irradiated Wheat," *Journal of Food Science*, 43:1102.

122 Frank, H. K., R. Münzner and J. F. Diehl. 1971. "Response of Toxigenic and Non-Toxigenic Strains of *Aspergillus flavus* to Irradiation," *Sabouraudia*, 9:21.

123 Webb, T. and T. Lang. 1987. *Food Irradiation: The Facts*. Wellingborough, Northhamptonshire: Thorsons Publishing Group.

124 Webb, T., T. Lang and K. Tucker. 1987. *Food Irradiation: Who Wants It?* Rochester, Vermont: Thorsons Publishers, Inc.

125 Frazer, J. G. 1974. *The Golden Bough (A Study in Magic and Religion)*. MacMillan Publishing Co., Inc.

126 Fafunso, M. and O. Bassir. 1976. "Effect of Cooking on the Vitamin C Content of Fresh Leaves and Wilted Leaves," *Journal of Agricultural and Food Chemistry*, 24:354.

127 FDA. 1986. "Irradiation in the Production, Processing, and Handling of Food," *Federal Register*, 51:13376.

128 Sugimura, T., M. Nagao, T. Kawachi, M. Honda, T. Yahagi, Y. Seino, S. Sato, N. Matsukura, T. Matsushima, A. Shirai, M. Sawamura and H. Mat-

sumoto. 1977. "Mutagen-Carcinogens in Foods with Special Reference to Highly Mutagenic Pyrolytic Products in Broiled Foods," in *Origins of Human Cancer, Book C*, H. H. Hiatt, J. D. Watson and J. A. Winsten, eds., New York: Cold Spring Harbour, pp. 1561–1576.

129 Sugimura, T. 1986. "The Formation of Mutagens and Carcinogens during Food Processing," in *Proceedings of the XIII International Congress of Nutrition*, T. G. Taylor and N. K. Jenkins, eds., London: John Libbey and Company, Ltd., pp. 833–837.

130 Sato, S. 1986. "Carcinogenicity of Mutagens Formed during Cooking," in *Proceedings of the XIII International Congress of Nutrition*, T. G. Taylor and N. K. Jenkins, eds., London: John Libbey and Company, Ltd., pp. 561–564.

131 Baskaram, C. and G. Sadasivan. 1975. "Effects of Feeding Irradiated Wheat to Malnourished Children," *American Journal of Clinical Nutrition*, 28:130.

132 Vijayalaxmi. 1976. "Genetic Effects of Feeding Irradiated Wheat to Mice," *Canadian Journal of Genetics and Cytology*, 18:231.

133 Vijayalaxmi and G. Sadasivan. 1975. "Chromosomal Aberration in Rats Fed Irradiated Wheat," *International Journal of Radiation Biology*, 27:135.

134 Vijayalaxmi. 1978. "Cytogenetic Studies in Monkeys Fed Irradiated Wheat," *Toxicology*, 9:181.

135 Bradsky, W. and I. V. Vryvaeva. 1977. "Cell Polyploidy: Its Relation to Tissue Growth and Function," *International Review of Cytology*, 50:275–332.

136 Armendaras, S., F. Salamanca and S. Frenk. 1971. "Chromosome Abnormalities in Severe Protein Calorie, Malnutrition," *Nature*, 232:271.

137 Vijayalaxmi and S. G. Srikantia. 1989. "A Review of the Studies on the Wholesomeness of Irradiated Wheat, Conducted at the National Institute of Nutrition, India," *Radiation and Physical Chemistry*, 34(6):941.

138 Sanderson, D. C. W. 1990. "Luminescence Detection of Irradiated Foods," in *Food Irradiation and the Chemist*, D. E. Johnston and M. H. Stevenson, eds., Conference, Belfast, Northern Ireland, April 1990, Royal Society of Chemistry, U.K.

139 Stevenson, M. H. and R. Grey. 1990. "Can ESR Spectroscopy Be Used to Detect Irradiated Foods," in *Food Irradiation and the Chemist*, D. E. Johnston and M. H. Stevenson, eds., Conference, Belfast, Northern Ireland, April 1990, Royal Society of Chemistry, U.K.

140 Raffi, J., M. H. Stevenson, M. Kent, J. M. Thiery and J. J. Belliardo. 1992. "European Intercomparison on Electron Spin Resonance Identification of Irradiated Foodstuffs," *International Journal of Food Science and Technology*, 27(2):111–124.

141 Diehl, J. F. 1991. "Nutritional Effects of Combining Irradiation with Other Treatments," *Food Control*, 2(1):20.

142 Kilcast, D. 1991. "Irradiation and Combination Treatments," *Food Control*, 2(1):6.

143 Food and Drug Administration. 1986. "Irradiation in the Production, Processing, and Handling of Food," *Federal Register*, 51:13376.

144 Kunstadt, P. 1989. "Transport and Disposal of Cobalt-60 Industrial Radiation Sources," *IAEA Technical Document*, 490:95.

145 Satin, M. 1977. "Is It 'Natural' or 'Nutritional'," *The Washington Post*, E3(Sept. 22).

146 Satin, M., B. McKeown and C. Findlay. 1978. "Design of a Commercial Natural Fibre White Bread," *Cereal Foods World*, 23(11):676.

147 Satin, M. 1980. "High Fibre White Bread," United States Patent, Number 4,237,170, filed on December 28, 1976, issued on December 2, 1980.

148 Satin, M. 1979. "High Fibre White Bread," Canadian Patent, Number 1,048,848, filed December 14, 1977, issued February 20, 1979.

149 Morris, M. E. 1981. "Whole White Bread Matches Nutrition of Whole Wheat," *Food Engineering*, 69(July).

150 Hughes, K. A. 1987. "You Would Think These Papayas Would Be Easy to Find at Night," *The Wall Street Journal* (March 31).

151 Dodson, M. 1987. "Small Group Pickets Irvine Supermarket Making Test Sales of Irradiated Papayas," *Los Angeles Times* (March 30).

152 Hall, T. 1987. "Food Industry Eyes Irradiation Warily," *New York Times* (April 1).

153 Titlebaum, L. F. and E. Z. Dubin. 1983. "Will Consumers Accept Irradiated Foods?" *Journal of Food Safety*, 5(4):219.

154 Bruhn, C. M., R. Sommer and H. G. Schutz. 1986. "Effect of an Educational Pamphlet and Posters on Attitude toward Food Irradiation," *Journal of Industrial Irradiation Technology*, 4(1):1.

155 Bruhn, C. M., H. G. Schutz and R. Sommer. 1988. "Food Irradiation and Consumer Values," *Ecology of Food and Nutrition*, 21:219.

156 Bruhn, C. M. and H. G. Schutz. 1989. "Consumer Awareness and Outlook for Acceptance of Food Irradiation," *Food Technology*, 43(7):93.

157 Malone, J. W., Jr. 1990. "Consumer Willingness to Purchase and to Pay More for Potential Benefits of Irradiated Fresh Food Products," *Agribusiness*, 6(2):163.

158 Anonymous. 1986. "Irradiated Fruit on Sale in Miami," *The New York Times* (Sunday, September 14).

159 Toufexis, A., J. M. Horowitz and R. Thompsom. 1986. "Food Fight over Gamma Rays," *Time Magazine* (22 September).

160 Phillips, D. 1986. "Irradiated Produce Hits Miami Market," *The Packer* (September 13).

161 Laurenzo, D., personal communication with author.

162 Harinasuta, T., M. Riganti and D. Bunnag, Faculty of Tropical Medicine, Mahidol University, Bangkok, Thailand, personal communication with author.

163 Waterfield, L. 1991. "Lining Up over Kilorads," *The Packer* (March 2).

164 McGuirk, A. M., W. P. Preston and A. McCormick. 1990. "Toward the Development of Marketing Strategies for Food Safety Attributes," *Agribusiness*, 6(4):297.

165 Bruhn, C. M. and J. Noell. 1987. "Consumer In-Store Response to Irradiated Papayas," *Food Technology*, 41(9):83.

166 Ingersoll, B. 1991. "Irradiation Foes Plan Media Blitz to Block Plant," *The Wall Street Journal*, B1(26 June).

167 Hecht, M. M. 1991. "Food and Water, Inc. – Turning Antinuclear Fear into Profits," *21st Century* (Fall):20–23.

168 1991. International Agency for Research on Cancer (WHO) IARC Monographs on the Evaluation of Carcinogenic Risks to Humans, Volume 52, Chlorinated Drinking Water; Chlorination By-Products; Some Other Halogenated Compounds; Cobalt and Cobalt Compounds, Lyon, France.

169 Hecht, M. M. 1991. "First U.S. Food Irradiation Plant Opens in Florida," *21st Century* (Fall):12–24.

170 Anonymous. 1991. "Beware Scares about Irradiated Food Risks," *The Christian Science Monitor* (17 July):13.

171 Ferguson, T. W. 1991. "Southern Fried Shipper Would Blast the Buffet Bugs," *The Wall Street Journal*, 10(September):A21.

172 Stossel, J. 1991. "The Power of Fear–An ABC TV 20/20 News Special" (13 December).

173 Booth, W. 1992. "U.S. Widens Food Irradiation," *International Herald Tribune*, 3(13 January).

174 Begley, S. and E. Roberts. 1992. "Dishing Up Gamma Rays," *Newsweek* (27 January):50–51.

175 Rubin, S. L. 1992. "Lake Region Rotarians Sample Sweet Irradiated Strawberries," *Haines City Herald* (31 January).

176 Bauser, J. E. 1992. "Food Irradiation Is Safe and Good," *Naples Daily News* (9 February).

177 Standard, A. C. 1992. "Dispelling the Myths about Food Irradiation," *The Packer* (8 February).

178 Marcotte, M. 1992. "Irradiated Strawberries Enter the U.S. Market," *Food Technology*, 46(5):80–86.

179 Pratt, S. 1992. "Nuked Fruit–Grocer Bets on Irradiation to Enhance Produce," *Chicago Tribune*, 7(3), 19 March.

180 Aschoff, S. 1992. "On a Mission to Zap Your Chicken," *Florida Trend* (March):52–56.

181 Conley, S. T. 1992. "'What Do Consumers Think about Irradiated Foods? Food Safety and Inspection Service," *Food Safety Review* (Fall).

182 Mason, J. D. 1992. "Food Irradiation–Promising Technology for Public Health," *Public Health Reports*, 107(5):489.

183 Engel, R. E. Food Safety and Inspection Service, U.S.D.A., Washington, D.C., personal communication.

184 Memorandum from the Chairperson, Foodborne and Waterborne Diseases Prevention Working Group of the Center for Disease Control, Atlanta, Georgia, to Members of the Working Group, March 12, 1993.

185 Pszczola, D. E. 1992. "Irradiated Produce Reaches Midwest Market," *Food Technology*, 46(5):89–92.

186 Leemhorst, J. G. 1994. Personal communication.

187 Radomyski, T., E. A. Murano and D. G. Olson. 1993. "Irradiation of Meat Products to Ensure Hygienic Quality," *Dairy, Food and Environmental Sanitation*, 13(7):398–403.

188 Kwon, J. H., M. W. Byun, S. B. Warrier, A. S. Kamat, M. D. Alur and P. M. Nair. 1993. "Quality Changes in Irradiated and Non-Irradiated Boiled-Dried Anchovies after Inter-Country Transportation and Storage at 25 Degrees," *Journal of Food Science and Technology—India*, 30(4):256–260.

189 Clavero, M. R. S., J. D. Monk, L. R. Beuchat, M. P. Doyle and R. E. Brackett. 1994. "Inactivation of *Escherichia coli* O157:H7, *Salmonellae*, and *Campylobacter jejuni* in Raw Ground Beef by Gamma-Irradiation," *Applied and Environmental Microbiology*, 60(6):2069–2075.

190 Ama, A. A., M. K. Hamdy and R. T. Toledo. 1994. "Effects of Heating, pH and Thermoradiation on Inactivation of *Vibrio vulnificans*," *Food Microbiology*, 11(3):215–227.

191 Rashid, H. O., H. Ito and I. Ishigaki. 1992. "Distribution of Pathogenic Vibrios and Other Bacteria in Imported Frozen Shrimps and Their Decontamination by Gamma Irradiation," *World Journal of Microbiology and Biotechnology*, 8(5):494–499.

192 Thayer, D. W. 1994. "Wholesomeness of Irradiated Foods," *Food Technology*, 48(5):132–136.

193 Rodriguez, H. R., J. A. Lasta, R. A. Mallo and N. Marchevsky. 1993. "Low-Dose Gamma Irradiation and Refrigeration to Extend Shelf Life of Aerobically Packed Fresh Beef Round," *Journal of Food Protection*, 56(6):505–509.

194 Naik, G. N., P. Paul, S. P. Chawla, A. T. Sherikar and P. M. Nair. 1993. "Improvement in Microbiological Quality and Shelf-Life of Buffalo Meat at Ambient Temperature by Gamma-Irradiation," *Journal of Food Safety*, 13(3):177–183.

195 Frencia, J. P. 1992. "Prolongation of the Shelf Life of Offal by Irradiation," *Viandes et Produits Carnes*, 13(3):69–74.

196 Thayer, D. W., G. Boyd and R. K. Jenkins. 1993. "Low-Dose Gamma Irradiation and Refrigerated Storage in Vacuo Affect Microbial Flora of Fresh Pork," *Journal of Food Science*, 58(4):717–719.

197 Farkas, J. and E. Andrassy. 1993. "Interaction of Ionizing Radiation and

Acidulants on the Growth of the Microflora of Vacuum-Packaged Chilled Meat Product," *International Journal of Food Microbiology*, 19(2):145–152.

198 Grandison, A. S. and A. Jennings. 1993. "Extension of the Shelf Life of Fresh Minced Chicken Meat by Electron Beam Irradiation Combined with Modified Atmosphere Packaging," *Food Control*, 4(2):83–88.

199 Moral-Rama, A. 1993. "Refrigeration and Preservation of Fish Packaged under Modified Atmosphere for Retail Sale," *Alimentacion Equipos y Technologia*, 12(1):101–104.

200 Lambert, A. D., J. P. Smith and K. L. Dodds. 1992. "Physical, Chemical and Sensory Changes in Irradiated Fresh Pork Packaged in Modified Atmosphere," *Journal of Food Science*, 57(6):1294–1299.

201 Wessels, M. L. and Plessis-Adu. 1992. "Gamma Radurisation of Vienna Sausages," *Food Industries*, 45(6):14–16.

202 Lambert, A. D., J. P. Smith, K. L. Dodds and R. Charbonneau. 1992. "Microbiological Changes and Shelf Life of MAP-Irradiated Fresh Pork," *Food Microbiology*, 9(3):231–244.

203 Shamsuzzaman, K., N. Chuaqui-Offermans, L. Lucht, T. McDougall and J. Borsa. 1992. "Microbiological and Other Characteristics of Chicken Breast Meat Following Electron-Beam and Sous-Vide Treatments," *Journal of Food Protection*, 55(7):528–533.

204 Patterson, M. F., A. P. Damoglou and R. K. Buick. 1993. "Effects of Irradiation Dose and Storage Temperature on the Growth of *Listeria monocytogenes* on Poultry Meat," *Food Microbiology*, 10(3):197–203.

205 Kovacs, A., P. Hargittai, L. Kaszanyiczki and G. Foeldiak. 1994. "Evaluation of Multipurpose Electron Irradiation of Packaged and Bulk Spices," *Applied Radiation and Isotopes*, 45(7):783–788.

206 Katusin-Razem, B., S. Duric-Bezmalinovic, D. Razem, S. Matic, V. Mihokovic and A. Dunaj. 1993. "Microbial Status of Dry Soup Greens and Vegetable Seasonings. Decontamination of Dehydrated Leek by Irradiation," *Prehrambeno Tehnoloska i Biotehnoloska Revija*, 30(4):165–170.

207 Delincee, H. and A. Bogner. 1993. "Effect of Ionizing Radiation on the Nutritional Value of Legumes," *Bioavailability '93 – Nutritional, Chemical and Food Processing Implications of Nutrient Availability (Federation of European Chemical Societies Nutrient Bioavailability Symposium)*, 2:267–371.

208 Roy, M. K. and H. H. Prasad. 1993. "Gamma Radiation in the Control of Important Storage Pests of Three Grain Legumes," *Journal of Food Science and Technology – India*, 30:(4)275–278.

209 Miller, W. R., E. J. Mitcham, R. E. McDonald and J. R. King. 1994. "Postharvest Storage Quality of Gamma-Irradiated Climax Rabbiteye Blueberries," *HortScience*, 29(2):98–101.

210 Sattar, A., Neelofar and M. A. Akhtar. 1992. "Radiation Effect on Ascorbic

Acid and Riboflavin Biosynthesis in Germinating Soybean,'' *Plant Foods for Human Nutrition*, 42(4):305–312.

211 Armstrong, S. G., S. G. Wyllie and D. N. Leach. 1994. "Effects of Preservation by Gamma-Irradiation on the Nutritional Quality of Australian Fish,'' *Food Chemistry*, 50(4):351–357.

212 Damayanti, M., G. J. Sharma and S. C. Kundu. 1992. "Gamma Radiation Influences Postharvest Disease Incidence of Pineapple Fruit,'' *HortScience*, 27(7):807–808.

213 Hasan, A. J. K., N. Choudury, A. Begum and N. Nahar. 1993. "Preservation of Vegetables by Microbial Activity and Irradiation,'' *World Journal of Microbiology and Biotechnology*, 9(1):73–76.

214 Shamsuzzaman, K. and L. Lucht. 1993. Resistance of *Clostridium sporogenes* spores to Radiation and Heat in Various Nonaqueous Suspension Media, *Journal of Food Protection*, 56(1):10–12.

215 Borsa, J., W. S. Chelak, R. R. Marquardt and A. A. Frohlich. 1992. "Comparison of Irradiation and Chemical Fumigation Used in Grain Disinfestation on Production of Ochratoxin A by *Aspergillus alutaceus* in Treated Barley,'' *Journal of Food Protection*, 55(12):990–994.

216 Pinn, A. B. O., C. Colli and J. Mancini-Filho. 1993. "Beans (*Phaseolis vulgaris* L.) Irradiation–Iron Availability,'' *Bioavailability '93–Nutritional, Chemical and Food Processing Implications of Nutrient Availability (Federation of European Chemical Societies Nutrient Bioavailability Symposium)*, 2:165–199.

217 Pfeiffer, C., F. Diehl and W. Schwack. 1993. "Effect of Irradiation on Bioavailability of Food Folates,'' *Bioavailability '93–Nutritional, Chemical and Food Processing Implications of Nutrient Availability (Federation of European Chemical Societies Nutrient Bioavailability Symposium)*, 2:408–411.

218 Mitchell, G. E., R. L. McLauchlan, A. R. Issacs, D. J. Williams, and S. M. Nottingham. 1992. "Effect of Low Dose Irradiation on Composition of Tropical Fruits and Vegetables,'' *Journal of Food Composition and Analysis*, 5(4):291–311.

219 Reineccius, G. A. 1992. "Irradiation and Flavor,'' *Cereal Foods World*, 37(7):505.

220 Grant, I. R., C. R. Nixon and M. F. Patterson. 1993. "Effect of Low-Dose Irradiation on Growth of and Toxin Production by *Staphylococcus aureus* and *Bacillus cereus* in Roast Beef and Gravy,'' *International Journal of Food Microbiology*, 18(1):25–36.

221 Fuchs, E. and H. Heusinger. 1994. "Ultrasound-Induced Scission of the Glycosidic Bond in Disaccharides,'' *Zeitschrift fuer Lebensmittel-Untersuchung und -Forschung*, 198(6):486–490.

222 Portenlaenger, G. and H. Heusinger. 1992. "Chemical Reactions Induced by

Ultrasound and Gamma-Rays in Aqueous Solutions of L-Ascorbic Acid," *Carbohydrate Research*, 232(2):291−301.

223 McMurray, B. T., N. I. O. Brannigan, J. T. G. Hamilton, D. R. Boyd and M. H. Stevenson. 1994. "Detection of Irradiated Foods Using Cyclobutanones," *Food Science and Technology Today*, 8(2):99−100.

224 Stewart, E. M. and M. H. Stevenson. 1994. "The Use of Electron Spin Resonance (ESR) Spectroscopy for the Detection of Irradiated Crustacea," *Food Science and Technology Today*, 8(2):101−102.

225 Haine, H. E. and L. Jones. 1994. "Microgel Electrophoresis of DNA as a Method to Detect Irradiated Food," *Food Science and Technology Today*, 8(2):103−105.

226 Delincée, H. 1994. "Detection of Irradiated Foods Using Simple Screening Methods," *Food Science and Technology Today*, 8(2):109−110.

227 Goodman, B. A., W. Deighton and S.M. Glidewell. 1994. "Detection of Irradiated Foods of Plant Origin by Electron Paramagnetic Resonance Spectroscopy," *Food Science and Technology Today*, 8(2):110−111.

228 Hayashi, T., S. Todoriki, K. Otobe and K. Kohyama. 1993. "Applicability of an Impedance Method for Detection of Irradiation of Potatoes," *Journal of Japanese Society of Food Science and Technology*, 40(5):378−384.

229 Hayashi, T., S. Todoriki and K. Kohyama. 1993. "Applicability of Viscosity Measurement for Detection of Irradiation of Spices," *Journal of Japanese Society of Food Science and Technology*, 40(6):456−460.

230 Wirtanen, G., A. M. Sjoeberg, F. Boisen and T. Alanko. 1993. "Microbiological Screening Method for the Indication of Irradiation of Spices and Herbs: A BCR Collaborative Study," *Journal of the AOAC−International*, 76(3):674−681.

231 Hozova, B. and M. Takacsova. 1993. "The Influence of Combined Storage Procedures of Foods on B Vitamins Content Demonstrated at the Example of Heat Sterilization and Irradiation," *Nahrung*, 37(4):345−351.

232 Takacsova, M., S. Sorman, S. Dudasova, A. Rajniakova, B. Hozova and J. Slosiarova. 1992. "Changes in Characteristic Properties of Lipid Samples Preserved by Combination of Thermosterilization and Ionizing Radiation," *Potravinarske-Vedy*, 10(4):293−302.

233 Mosher, P. 1992. "Irradiation: in the Consumer's Hands," *Citrus and Vegetable Magazine*, 56(1):50−56.

234 Thayer, D. W. and G. Boyd. 1993. "Elimination of *Escherichia coli* O157:H7 in Meats by Gamma Irradiation," *Applied and Environmental Microbiology*, 59(4):1030−1034.

235 Dodds, K. L. 1993. "An Introduction to Predictive Microbiology and the Development and Use of Probability Models with *Clostridium botulinum*," *Journal of Industrial Microbiology*, 12(3/5):139−143.

236 *FAO/IAEA Consultants Meeting on Irradiation for Shelf-Stable Foods*, Vienna, Austria, 11−15 October, 1993.

237 Anonymous. 1994. "Consumer Group Asks FCC to Investigate ABC's Report on Food Irradiation," *Communications Daily* (February 23).

238 Anonymous. 1994. "FCC Investigation Sought of Possibel 20/20 New Distortion," *Food Chemical News* (January 17).

239 Anonymous. 1994. "Food & Hysteria," *The Wall Street Journal*, A14. April 27.

240 Katzenstein, L. 1992. "Food Irradiation," *American Health* (December):60–68.

241 Anonymous. 1993. "Good Food You Can't Get," *The Reader's Digest* (July).

242 Van Hazinga, C. 1993. "Irradiated Foods: Are They Safe?" *Better Homes and Gardens* (October):58–61.

243 Hecht, M. M. 1993. "Interview with Julia Child – Food Irradiation and Biotechnology Necessary to 'Protect Our People,' " *21st Century* (Spring):60–61.

244 Oprah Winfrey Television Show. 1994. "Bacteria in Foods Prepared at Home" (August 18).

245 Anonymous. 1994. "Espy and Daschle Unveil Legislation Targeting Microbial Contamination in Meat and Poultry," *USDA News* (September 14).

246 Espy, M. 1994. "Ensuring a Safer and Sounder Food Supply," *Food Technology* (September):91–93.

247 Anonymous. 1994. *The Kiplinger Agricultural Letter*, 65(5).

248 Anonymous. 1994. "Current Meat Inspection System 'Insufficient' for Ensuring Food Safety, Independent Panel Says," *American Gastroenterological Association Foundation News* (July 13).

249 Anonymous. 1994. "*E. coli* 157:H7 Infections: An Emerging National Health Crisis," *American Gastroenterological Association Foundation Consensus Conference Statement* (July 11–13).

250 Mason, J. O. 1992. "Food Irradiation – Promising Technology for Public Health," *Public Health Reports*, 107(5):489.

251 Lee, P. R. 1994. "From the Assistant Secretary for Health, US Public Health Service," *Journal of the American Medical Association*, 272(4):261.

252 Steele, J. H. 1993. "It's Time for Food Irradiation," *Journal of Public Health Policy*, 14(2):133–136.

253 Murphy, D. 1994. "Reconsidering a Hot Issue," *The National Provisioner* (February):8.

254 Vestal, D. 1994. "Has the Time Come for Irradiation?" *The Grower* (July):12–16.

255 Beghian, L., A. Brynjolfsson, F. J. Francis, G. G. Giddings, E. S. Josephson, M. Kroger, S. A. Miller, E. G. Remmers, J. H. Steele and E. M. Whelan. 1994. "Irradiation Is a Safe, Effective Way to Combat Food-Borne Pathogens," *Meat Marketing and Technology* (December):44.

256 American Meat Institute. 1993. "The More Consumers Learn about Irradiation, the More They Want It," *News Release* (October 7).

257 American Meat Institute. 1993. "AMI Foundation Finds Irradiation Kills Harmful Bacteria in Beef," *News Release* (October 7).

258 Beuchat, L. R., M. P. Doyle and R. E. Brackett. 1993. "Irradiation Inactivation of Bacterial Pathogens in Ground Beef," a study prepared for the American Meat Institute by the Center for Food Safety and Quality Enhancement, University of Georgia, Griffin, GA.

259 American Meat Institute. 1994. "Issue: Irradiation and Food Safety," issues briefing, October.

260 American Meat Institute. 1993. "Consumer Awareness, Knowledge, and Acceptance of Food Irradiation," a study conducted by the Gallup Organization and ABT Associates for the American Meat Institute, the Center for Food Safety and Quality Enhancement, University of Georgia, Griffin, GA.

261 Murphy, D. 1994. "The CEOs Speak," *The National Provisioner* (December):50−54.

262 Aylward, L. 1994. "Hormel's Johnson: Stop Paying Attention to the Flakes; Start Paying Attention to the Scientists," *Meat Marketing and Technology* (June):24−26.

263 Hyde, N. 1993. "Activists Preventing Poultry Industry from Detecting Harmful Bacteria," *Farm Bureau News* (May):8.

264 Jacoby, P., and J. Baller. 1994. "Let's Stop Playing Culinary Roulette and Get on with Irradiating Food," *Legal Opinion Letter*, 4(14).

265 Anonymous. 1994. *The Kiplinger Agricultural Letter*, 65(3).

266 Anonymous. 1994. "Pope John Paul II Sent a Strong Message Supporting Nuclear Energy," *EIRNS* (Reuters) (April 15).

267 Pszczola, D. 1993. "Irradiated Poultry Makes U.S. Debut in Midwest and Florida Markets," *Food Technology* (November):89−96.

268 Anonymous. 1992. "Human Error Blamed for Food Poisoning," *Food Manufacture* (December):14.

269 Anonymous. 1993. "New York Taco Bell Closes after Food Poisoning Outbreak," *Nation's Restaurant News* (July 5):2.

270 Anonymous. 1993. "Paprika Recall," *International Food Safety News*, 2(9):99.

271 Bussewitz, W. 1993. "Restaurant Requests Rise for Food Poisoning Cover," *National Underwriter Property & Casualty* (October 4):13.

272 Robeck, M. R. 1995. "Product Liability Issues Raised by the Growth in the Irradiated Food Market," presented at the *Houston Fresh Food Association and the University of Texas Health Sciences Center Conference, Food Irradiation: Issues and Challenges*, Houston, Texas, January 20, 1995.

273 Corrigan, J. P. 1993. "Experiences in Selling Irradiated Foods at the Retail Level," in *Cost-Benefit Aspects of Food Irradiation Processing, Proceedings*

of an International Symposium on Cost-Benefit Aspects of Food Irradiation Processing (organized by FAO, WHO, and IAEA), International Atomic Energy Agency, Vienna, pp. 447–454.

274 Anonymous. 1994. "Ground Beef Sampling Program Preliminary Injunction Sought by Industry Groups," *Food Chemical News*, 36(37):60.

275 Burdock, G. A. 1993. "Review of the Biological and Chemical Data on High Dose Irradiation of Food," a report prepared for the World Health Organization Food Safety Unit (October).

276 FAO/IAEA. 1993. Report of *Consultants' Meeting on Irradiation for Shelf-Stable Foods*, Vienna, Austria (11–15 October).

277 Smith, G. L. 1992. "Utilization of Enzymes to Provide Heating Endpoint Markers and Modify Endogenous Cholesterol in Muscle Foods," *Dissertation Abstracts International*, B52(9):4538–4539.

278 Ohmori, T., T. Shigehisa, S. Taji and R. Hayashi, 1991. *Agricultural and Biological Chemistry*, 55(2):357–361.

279 Mertens, B. 1993. "Developments in High Pressure Food Processing. I.," *Internationale Zeitschrift fuer Lebensmittel Technik, Marketing, Verpackung und Analytik*, 44(3):100–104.

280 *FAO/IAEA Consultants Meeting on Irradiation for Shelf-Stable Foods*, Vienna, Austria, 11–15 October, 1993.

281 Mans, J. and B. Swientek. 1993. "Electrifying Progress in Aseptic Technology," *Prepared Foods*, 162(9):151–152.

282 Reuter, H. 1993. *Aseptic Processing of Foods*, Lancaster, PA: Technomic Publishing Co., Inc.

283 Kalman, B. 1994. Agroster Irradiation, Inc., Budapest, Hungary, personal communication.

284 UNEP. 1991. *Handbook for the Montreal Protocol on Substances That Deplete the Ozone Layer, 2nd Ed.* United Nations Environmental Program.

285 Moretti, R. H., 1993. "Cost-Benefit of Poultry Irradiation in Brazil," in *Cost-Benefit Aspects of Food Irradiation Process, Proceedings of an International Symposium on Cost-Benefit Aspects of Food Irradiation Processing* (organized by FAO, WHO and IAEA), International Atomic Energy Agency, Vienna, pp. 291–300.

286 Giddings, G. G. 1994. "Approved Sources of Ionizing Radiation," petition to FDA by Isomedix Inc. (June 30).

287 Coble, Y. D. 1993. "Irradiation of Food," American Medical Association Report of the Council on Scientific Affairs.

288 Anonymous. 1994. "Medical Experts Urge Radiation of Beef to Kill Deadly Bacteria," *The New York Times National* (July 14).

289 Anonymous. 1994. "Food Experts Call on Government to Set up System to Irradiate Meat," *Washington Times* (July 14).

290 Manning, J. 1994. "Hamburger Tied to City Illnesses," *Milwaukee Sentinel* (July 14):1A.

291 Kidwell, D. 1994. "Boca Man, 82, Gets Food Poisoning; Pink Meat Blamed," *The Miami Herald* (August 11):2B.

292 Anonymous. 1994. "An Outbreak of *Salmonella* Poisoning," *The International Herald Tribune* (October 22):P3.

293 Brody, J. E. 1994. "Fears of Food Irradiation Are Greatly Exaggerated," *The New York Times* (October 12).

294 Hurtado, M. E. 1994. "Food Irradiation Little More Than an Expensive Quick Fix," *Financial Times* (October 26).

295 Jones, M. E. 1994. Personal communication from the Director of Public Relations, Publix Supermarkets.

296 Rowe, S.M. 1995. "Looking for Safe Food," *The Oregonian* (August 12).

INDEX

Morton Satin is a recognized authority on the development of creative and practical approaches to the food industry. Over the past twenty years he has invented several commercially viable products and processing technologies for developed and developing countries alike.

An expert in product development and marketing, Mr. Satin has devoted his career to the food and agricultural sector. He has directed the functions of research and development, marketing, business development and quality assurance for a number of large multinational food corporations.

Trained in Canada as a molecular biologist, Mr. Satin has produced numerous publications and articles on food and agricultural technology, both for the technical and lay press. He has won several awards for his contributions to the industry, and has served as a special consultant to several levels of government, expert committees, international organizations and trade associations. He has participated in several high-level international missions for the promotion of technology transfer and industrial development.

An entrepreneur by nature, Mr. Satin brings to his work a unique mixture of experiences from both the private and public sectors, as well as from his academic and writing activities.